ENGINEERING LIBRARY

D0848379

Human Tissue Research

A European Perspective on the Ethical and Legal Challenges

Human Tissue Research

A European Perspective on the Ethical and Legal Challenges

FIRST EDITION

Edited by

Christian Lenk

Department of Medical Ethics and History of Medicine,
Goettingen University Medical Center, Germany

Nils Hoppe

CELLS – Centre for Ethics and Law in the Life Sciences,
Leibniz Universitaet Hannover, Germany

Katharina Beier

Department of Medical Ethics and History of Medicine,
Goettingen University Medical Center, Germany

Claudia Wiesemann

Department of Medical Ethics and History of Medicine,
Goettingen University Medical Center, Germany

OXFORD
UNIVERSITY PRESS

OXFORD
UNIVERSITY PRESS

Great Clarendon Street, Oxford OX2 6DP

Oxford University Press is a department of the University of Oxford.
It furthers the University's objective of excellence in research, scholarship,
and education by publishing worldwide in

Oxford New York

Auckland Cape Town Dar es Salaam Hong Kong Karachi
Kuala Lumpur Madrid Melbourne Mexico City Nairobi
New Delhi Shanghai Taipei Toronto

With offices in

Argentina Austria Brazil Chile Czech Republic France Greece
Guatemala Hungary Italy Japan Poland Portugal Singapore
South Korea Switzerland Thailand Turkey Ukraine Vietnam

Oxford is a registered trade mark of Oxford University Press
in the UK and in certain other countries

Published in the United States
by Oxford University Press Inc., New York

First published 2011

British Library Cataloguing in Publication Data

Data available

Library of Congress Cataloging in Publication Data

Data available

Typeset in Minion by Glyph International, Bangalore, India
Printed in Great Britain
on acid-free paper by
CPI Anthony Rowe, Chippenham, Wiltshire

ISBN 978–0–19–958755–1

10 9 8 7 6 5 4 3 2 1

Contents

Preface

Human tissue has become an increasingly important raw material for medical research and scientific progress. In continuation of the medical scientific tradition to gather rare samples of human tissues, organs, and body parts, body materials of patients as well as healthy research participants are now collected in large repositories and biobanks. It is the scientific aim to identify connections as yet unknown between the human genome and multifactorial diseases. Research results may be especially useful for pharmaceutical companies who themselves operate large biobanks. Therefore body material of specific patient groups is assigned a particularly high scientific as well as financial value.

Because of a number of high-profile scandals in the UK, the public debate in relation to organ and tissue storage gathered fresh momentum in the late 1990s, carrying it away from the state of play in academic and political debate. At the Royal Liverpool Children's NHS Trust and the Bristol Royal Infirmary, large quantities of human cadaveric tissue and organs were retained without adequate consent after post-mortem examinations. More recently, in 2006, fresh storage incidents prompted the publication of the controversial Madden Report on post-mortem practice and procedures in Irish hospitals. The routine harvesting of tissue and cells for therapeutic purposes in the USA, which were then imported into Germany as a medical product, led to widespread fears of contamination of tissue-based medical products by Creutzfeldt–Jakob disease.

Whilst the ethical discussion concerning the use of human tissue for therapeutic purposes is seemingly advanced in Europe and in consequence of a EC Directive is already subject to national legislation, analysis of the ethical and legal regulation of research with human tissue has received relatively little attention. Taking into account the fact that the procurement, storage, and transfer of human tissue for research purposes raise questions of their own, in the present volume we tackle these issues by means of a thorough ethical, philosophical, and legal investigation. As tissue-related research tends to be most fruitful if samples and information are shared across national borders, the heterogeneity of current rules and guidelines within the member states of the European Union demands further clarification. For this purpose, the present book connects ethical and legal considerations in the field of human tissue research to other important fields such as tissue engineering and research with human stem cells. By this means, the most important similarities and differences in the national ethical and legal frameworks will be elaborated.

The underlying rationale for investigating different approaches to tissue research and their respective ethical and legal implications results from the insight that the desired positive effects of human tissue research can only be achieved if the risks to individual autonomy, privacy, and justice are confronted adequately. Taking these challenges head-on, the present volume contributes to the discussion and development of appropriate ethical and legal guidelines from both an interdisciplinary and a comparative perspective. By delineating their respective national experiences and assessing current practice in tissue research, authors from different disciplines, such as bioethics, political sciences, law, sociology, and pathology, illustrate the wide range of ethical deliberation in this rapidly expanding field. At the same time, the evaluation of different regulatory models provides a basis for some normative conclusions concerning the most urgent requirements for biobanking in the European arena. Thus this book will contribute to a Europe-wide convergence and harmonization of policies for tissue-related research.

The concept of systematic computer-based large-scale research with human tissue in disease-related or epidemiological biobanks is one of the most recent and innovative developments in the field. Although human tissue research comprises a wider array than biobanking alone, many of the contributors to this volume deal with the latter because biobanks provide an instructive example for illustrating the major ethical and legal challenges that are pertinent to human tissue research in general. The special characteristics which make biobanks important for ethical and legal analysis include combination of tissue research with genetic analysis, reference to specific groups of patients or probands, and systematic and regular exchange of samples and data with third parties who have no direct contact to the patient. Other themes in the present volume, such as respect for human dignity and the question of property rights in human tissue, are more fundamental and are as applicable to tissue research in general as they are to biobank research in particular.

In this context it must be noted that the term 'biobank' has a variety of meanings. This terminological heterogeneity is partly due to different national approaches in research and research politics, but is also triggered by the different requirements and preconditions of diverse medical disciplines. Both aspects of this terminological heterogeneity are reflected in this volume. For example, existing tissue collections and their possible transformation into a biobank play a more important role for the pathologist than for the cardiologist who is interested in establishing a totally new tissue collection that can be used to carry out research on children with congenital heart failure. Thus it follows that the definition of the concept of (research) biobanks is directly affected by the specific circumstances of its establishment. By taking the meaning of 'bank' literally, some researchers argue that biobanks should not only store tissue samples, but also distribute them to researchers and exchange them with other biobanks. However, at present not all the local tissue collections referred to as 'biobanks' may satisfy this criterion.

In its statement *Biobanks for Research*, the German National Ethics Council (now the German Ethics Council) used a rather broad definition by stipulating that biobanks are 'collections of samples of human body substances which are or can be linked to person-related data and information from the donors'.[1] This means that a mere collection of body materials as such is not a biobank. To meet this definition, it is additionally required that existing samples can be connected with person-related data, although it is not explicitly stated that these data have to stem from genetic analysis (which is indeed the case in most biobank projects).

The Council of Europe's *Recommendation Rec (2006)4 of the Committee of Ministers to Member States on Research on Biological Materials of Human Origin*[2] gives a more detailed definition of epidemiological or population biobanks (Chapter V, Art. 17). Following this definition, such biobanks should show the following characteristics:

(1) the collection has a population basis;

(2) it is established, or has been converted, to supply biological materials or data derived therefrom for multiple future research projects;

---

[1] **Nationaler Ethikrat (German National Ethics Council)** (2004). *Biobanken für die Forschung. Stellungnahme (Biobanks for Research)*. Available at: www.ethikrat.org/dateien/pdf/Stellungnahme_Biobanken.pdf/ (accessed 20 January 2010), 24.

[2] **Council of Europe** (2006). *Recommendation Rec(2006)4 of the Committee of Ministers to Member States on Research on Biological Materials of Human Origin*. Available at http://www.coe.int/t/dg3/healthbioethic/texts_and_documents/Rec_2006_4.pdf (accessed 11 May 2010).

(3) it contains biological materials and associated personal data, which may include or be linked to genealogical, medical, and lifestyle data, and which may be regularly updated;

(4) it receives and supplies materials in an organized manner.

This description fits the large epidemiological biobank projects like the UK Biobank or deCODE Genetics in Iceland. As is further recommended in this document, biobanks which adhere to this definition 'should be subject to independent oversight' (Art. 19). Indeed some European countries demand an official registration of biobanks at the responsible ministries. Such national demands and requirements lead to a homogenization of the understanding of the term 'biobank'. However, any ethical analysis of biobanks has to cope with the wide range of functions and usage established to date.

An overview of the book's structure and the individual chapters is provided below. This volume combines (1) important ethical and legal themes and principles with (2) detailed descriptions of the situation in different countries and (3) contributions from different scientific disciplines which are important and influential in the field of tissue research. The contributions in Part I deal with fundamental ethical concepts, such as human dignity, respect for persons, non-maleficence, solidarity, reciprocity, and trust, which are applied to the field of human tissue research. In Part II, central legal approaches relevant to tissue-related research and biobanking are introduced and discussed. The various themes comprise the relationship of ethics and law, the legal status of human tissue, examples for different modes of regulation from three European countries (Switzerland, France, and Italy), and the problem of the intra-European transfer of tissue samples for research purposes. Additionally, in other chapters one will also find detailed information on the ethical and legal situation in Austria (Chapter 17), Germany (Chapters 6, 14, and 16), Sweden and Norway (Chapter 7), and the UK (Chapters 3, 8, and 16).

In Part III, research using human tissue is approached from an interdisciplinary perspective by including contributions from pathology, ethics, and sociology. The various themes in this part take up questions of clinical practice, informed consent in the case of human tissue engineered products, and the perception of informed consent by patients and medical staff.

Chapter summaries

The right to decide on and to control one's own body and the cultural framework of how to deal with the human body are at the centre of Michael Barilan's contribution (Chapter 1). The author makes it clear that he regards the current discussion on property in body material and the control of the use of body substances as a footnote to the history of anatomy and pathology in particular, and broader cultural history in general. He demonstrates that although we generally assume that our handling of the living or deceased body and its parts is perfectly rational, on the contrary it is often inconsistent with other important ethical and legal norms. For example, while the law does not forbid a person to put him/herself in danger of being eaten by wild animals, it does not allow the corpse of this person—even against their expressed will—to be fed to the same animals. On the other hand, it is a special characteristic of Western medicine and ethics that the use of human remains for scientific, religious, and educational purposes is often permitted. Although 'undignified' activities with the human corpse are usually prohibited by the legal system, the use of body materials (e.g. for scientific analysis) is widely perceived as compatible with human dignity, especially when it is based on the valid informed consent of the person. Thus, in his conclusion, Barilan argues for a well-balanced compromise between the interests of the living and the dignity of the dead.

In Chapter 2 the Dutch philosophers Rieke van der Graaf and Johannes van Delden consider whether the established principles of informed consent are of any value in the context of biobanks.

They identify four scenarios where informed consent is problematic. First, participants of biobank research can only ever be broadly informed about the research to be undertaken with their samples. Secondly, the widespread practice of resorting to the use of material left over from diagnostics or surgery means that, in most cases, the subsequent research is not covered by the original consent for the diagnostic or surgical intervention. Thirdly, the authors question the validity of the debate surrounding a practice of recontacting participants in biobank research in order to obtain 'fresh' informed consent for each and every research undertaking. Finally, they argue that the right to withdraw consent for participation in biobank research is problematic as it may adversely affect the quality of the research. Instead, van der Graaf and van Delden suggest that it is permissible for biobank research to be covered by a broad consent on the basis that participants have sufficient reasons to consent to the research and that they share the same ends as the researchers. They argue that the guiding principle is that participants in biobank research must not be used merely as a means to an end.

In Chapter 3 Austen Garwood-Gowers analyses the key implications of respect for the individual regarding the discourse, practice, and governance of human tissue research. By 'respect' he understands the Kantian principle of never treating people as a mere means to an end, showing that this aspect is a major normative requirement in different international and national documents. According to Garwood-Gowers, respect provides the basis for a robust approach to information and consent standards in the field of human tissue research. However, by taking a closer look at the UK and other domestic regulations of human tissue research, he finds several deviations from a strict consent and information rule which appear incompatible with respect for the individual. Thus Garwood-Gowers concludes that the public should never take for granted the commitment of the regulating authorities to a responsible guardianship, including the protection of respect for the individual.

Nils Hoppe's contribution (Chapter 4) is devoted to the notion of risk. It is inherent in the research methodology of many biobanks that linkage to a specific individual is required for the research to be conducted and that the identifiability of the data subject hinges on the density of other data and contemporary technological methods. For the purposes of his chapter, he takes the position that these data cannot be anonymized appropriately and permanently, thus creating a particular type of risk for the donor. Hoppe investigates risk assessment and shows how the notion of risk differs in two normative functions: the ethical assessment of biobank research and the regulatory appraisal of such research. Furthermore, he points out certain difficulties with risk assessment itself, such as problems with balancing risk against benefit, as the risk, unlike the benefit, is solely to the individual participant. Other factors, such as monetary incentives or hope, also influence personal risk affinity. Thus Hoppe shows that risk is by no means as easy to assess as it seems.

In Chapter 5, Nadja Kanellopoulou examines the relationship between reciprocity, trust, and public interest for biobank research. How can we balance participants' interests against the demands of research and important considerations of social responsibility? The 'current regulatory emphasis on altruism in the UK' serves as a specific starting point in her work and the notion of altruism in the medical field serves as a more general context. She argues that it is one of the peculiarities of biobanks and genomic research that large numbers of volunteers are needed, which makes considerations regarding altruism and reciprocity far more important than in other fields of research. In this context, Kanellopoulou also addresses the question of how we should deal with genetic health information which can affect not only individual subjects but also their families as collective participants in biobank research. Patient interest groups and other forms of patient collectives make it appear to be natural to perceive the researcher–participant relationship from a social contract perspective. The conceptual framework of altruism, which is applied in the field of blood donation for example, has some obvious limitations which exacerbate a transfer to the area of tissue research and biobanking. In contrast, a reciprocal paradigm of tissue research seems to be a valuable alternative for future organization of the field.

In Chapter 6 Christian Lenk analyses the notion of solidarity in the context of biobank research in the special context of incidental health findings. The genetic screening of large amounts of tissue samples may well lead to important findings for hitherto hidden disease dispositions on the side of the tissue donor. However, at present there is no common standard in biobank research to inform participants about incidental health findings. As Lenk shows, a number of biobanks make extensive reference to altruism and solidarity in their self-portrayal to the broader public. Given that solidarity is not a one-sided but a mutual ethical obligation, not informing the donor of important health findings can be interpreted as an avoidable harm and may be seen as negligent in jurisdictions which have enshrined civil and criminal sanctions for failures to assist, such as the German jurisdiction. Lenk concludes that every biobank research framework that is based on the notion of solidarity should also actively support the interests of its participants. Informing patients of incidental health findings appears to be an important step towards an appropriate regulation of biobank research.

As regards the alignment of donors' and researchers' interests, in Chapter 7 Katharina Beier discusses two instructive examples of tissue research regulation in Sweden and Norway. In her analysis of these countries' biobank acts, Beier aims to leave behind the dichotomy of individualism versus communitarianism in the ethics of tissue research. In the Scandinavian regulatory framework, she identifies a conjunction of both perspectives: the wish to enable biobank research and thereby serving the common good, and the wish for the protection of individual interests. In order to provide an explanation of this feature, Beier highlights two concepts that reconcile these contrasting attitudes. Whilst the Swedish welfare state supports so-called *state individualism* which is based on a reciprocal relationship between the state and the individual citizen, the Norwegian *dugnad* tradition strengthens reciprocal assistance amongst its citizens. Both traditions can be fruitfully applied to the ethics of biobank research.

The history of organ and tissue collection and the Human Tissue Act 2004 in England and Wales is the starting point for Chapter 8 by José Miola. As he points out, the British organ retention scandal was partly due to the unsatisfactory legal regulation of the Human Tissue Act 1961. According to Miola, the autonomy of the donors and the accountability of the medical personnel are the two main principles of the Human Tissue Act 2004. Violations of the Act are now be treated as criminal acts with fines and/or imprisonment for up to three years. Thus the development and final form of the Human Tissue Act 2004 can be seen not only as a legal activity, but at the same time must be perceived as an integration of ethical and legal demands. According to Miola, this is a new development in medical ethics and law which should be welcomed by the stakeholders in the field.

The 'no-property rule' as a traditional axiom of common law is the main theme of Chapter 9 by Remigius Nwabueze. Although this rule has been influential not only in the legal systems of the UK but also in most other Western legal systems, there are significant exceptions from this principle, for example regarding the commodification of body parts through investment in work and skill. Thus, according to Nwabueze, we must ask whether the whole 'no-property rule' is still compatible with contemporary medical and technological demands and requirements. This question leads to a careful and detailed discussion of the advantages and disadvantages of the theoretical and pragmatic implications of this rule. In his conclusion, Nwabueze argues for a more complex legal approach to property and human tissue than the traditional no-property principle to do justice to current medical and scientific developments.

In Chapter10 Bianka Dörr describes the legal regulation of research using human biological material and personal data in Switzerland. As there is presently no such law in force, Dörr refers to the draft bill concerning the act on research with humans as well as to the relevant guidelines of the Swiss Academy of Medical Sciences (SAMS). When comparing the two documents, Dörr discusses various schemes of informed consent as well as different ethical regulations relating to data protection. In particular, she demonstrates that the aim of both documents is to ensure the protection of the participating individuals while, at the same time, providing optimized research conditions for the use of the material and data gathered. The Swiss legislator and the SAMS agree on the need to provide greater legal certainty and transparency in research which, in turn, may help to enhance public trust in research with humans and in science.

An analysis of the French regulation of research biobanks is provided by Virginie Commin in Chapter 11. In addition to the 2004 Bioethics Act, the Public Health Code and the Civil Code are pertinent in this field. Although the French legal framework already covers a broad range of issues that are important for biobanks, some ethical problems still remain unresolved. Commin particularly identifies the issues of access to biobanks and ownership. Whilst a coherent regulation for the former is lacking, the law remains silent on the latter question. Another question concerns the donor's user rights to his/her samples and related data. Commin highlights several shortcomings as regards the donor's information and consent about secondary use as well as the information about results from research with biobanks. Thus she concludes that the law should take more serious account of the collective dimension of participation in biobanking activities at both the national and international level.

The ethical and legal aspects of the collection and storage of tissue for research in Italy are addressed by Antonio Spagnolo, Viviana Daloiso, and Lara Parente in Chapter 12. As these authors show, an extensive scientific discussion has already taken place in Italy, leading to the publication of guidelines by several institutions. As well as analysing these documents, the authors also demonstrate the concordance of Italian guidelines with parts of the European recommendations, such as the *Recommendation No. R(94)1* of the Council of Europe. Equally, they expect an effect of guidance and homogenization by the existing national guidelines for the legal situation in Italy. Italy's ratification of the Oviedo Convention is interpreted as a general strengthening of the principle of informed consent which also affects the field of tissue research and biobanking. The authors conclude that biobanks are a very valuable resource for science in particular, and for society in general, but that their use has to be balanced very carefully against the rights of the individual donor.

In Chapter 13 Jasper Bovenberg maps the fundamental notion of freedom of movement in the European Union in the context of human tissues and cells. He convincingly argues that a high level of uncertainty and anxiety surrounds the issue of cross-border transfers of human biological material. In the absence of supranational legislation, he develops an approach on the basis of the regulation applied to the international transfer of personal data. This analogy permits the development of a number of practical legal mechanisms for ensuring appropriate protective systems for stakeholders: whilst cross-border shipments of tissue and cells are permitted, they are subject to EU-approved standard contractual clauses resulting in uniformly enforceable civil remedies. This approach represents a regulatory shift from the public to the private sphere, resulting in flexibility, equity, and instant practical utility. Bovenberg's approach is an example of adapting existing norms to a new problem, maintaining the freedom of movement within the European Union and reinforcing the European Research Area, and represents a valuable contribution to the discussion surrounding the transfer of human tissues and cells.

In Chapter 14 Brochhausen and his colleagues address an often neglected field of tissue research—research with tissue archived in pathology institutes for diagnostic purposes. As there is currently no specific law regarding the use and misuse of archived tissue samples in Germany, they discuss and compare European guidelines and specific German regulations. In accordance with these guidelines, the authors stress the importance of a formalized informed consent procedure to protect the personal rights and freedom of tissue donors. To this end, they outline key elements for an informed consent form regarding research with human biological material. In particular, they argue for including transparent information about the scientific use of the material, allowing the donor to limit the extent of research purposes.

Medicinal products resulting from human tissue engineering are at the focus of Chapter 15 by Leen Trommelmans, Joseph Selling, and Kris Dierickx. The cells which serve as the raw material for such processes come from a number of different sources, among them ethically sensitive materials such as human embryonic stem cells, bone marrow, or umbilical cord blood. The ethical considerations in this field of research raise not only the question of physical harm, but most importantly possible non-physical risks and dangers regarding the autonomy and privacy of the donor or commercial exploitation. In addition, the high level of requests for tissue and cells for research and therapy demonstrates that currently these are coveted materials. In particular, the authors argue that human tissue and cells are associated with various values. However, the value of these materials contrasts with the fact that the tissue source—the donor—will, as a default rule, not receive any financial reward for his/her donation. Trommelmans and colleagues see this as a potential point of contention between tissue donors and tissue users, as can be observed in other areas of tissue research such as the classic 'Moore case'.

In Chapter 16 Susanne Weber, Dana Wilson-Kovacs, and Christine Hauskeller focus on a national comparison of clinical trials involving autologous stem cells in heart repair in Germany and the UK. Although the clinical trials in question are subject to the regulatory framework of the EU, which has been established in order to harmonize the conditions for medical applications using tissues and cells throughout the member states, the authors succeed in showing a considerable impact of national and international governance practices on stem-cell research in Germany and the UK. Weber and her colleagues describe the integration of European regulations on medicinal products, tissues, and cells into German and UK regulatory processes and institutions by focusing particularly on how biological risks are negotiated in each country. Finally, the authors discuss the implications of their findings for the harmonization of regulations involving novel cell therapies and identify further directions of research.

The specific functions and dynamics of informed consent in the field of human tissue research are at the centre of Chapter 17 by Milena Bister. She analyses the results of an Austrian case study which examined patients' and researchers' perceptions of the practice of giving consent from a sociological perspective. While patients awaiting surgery show a permanent 'readiness of address', for the researcher informed consent represents the key to gain access to tissue as a research material in a legitimate way. Against this background, Bister argues for a re-conceptualization of informed consent as a purposive institutional procedure which structures the interaction and communication of patients and researchers. She concludes that the meaning of informed consent goes beyond the maintenance of individual autonomy rights and the researchers' protection against potential litigation by serving a variety of purposes on both sides.

Part I: Key concepts of the ethical debate

Human dignity and respect

This volume takes its starting point from considerations of fundamental ethical principles. Two of these principles are human dignity and respect for persons. Although in recent decades the concept of human dignity has been subject to criticism from a utilitarian and pragmatic point of view, it still serves—at least in the European setting—as a basic principle in the European Convention and is also referred to in national conventions and legislation. In addition, there appears to be a new willingness in the ethical discussion to analyse the notion of human dignity and to apply it to problems of biomedical research. Some theoretical concepts bind human dignity to autonomy as an existing and actual function of the human mind. However, this understanding leaves open whether human dignity also has to be assigned to human beings who do not have or have lost the ability for autonomous judgements and decisions. In contrast with such a rationalist and mentalist understanding of human dignity, the current interpretation sees it rather as a fundamental principle which can also explain and help to make decisions on bodily and mental states where autonomous choices of the person are no longer possible. This is a characteristic which the principle of human dignity definitely shares with the concept of respect for persons and their decisions. We respect the person and his/her once met decisions even if he/she is temporarily or persistently not in the position to decide for him/herself. Moreover, human dignity refers to the human body as the medium of personal identity and individuality as well as a manifestation of humankind. The importance of these findings for tissue research lies in the fact that a considerable number of tissue samples are taken from deceased persons, although this field is, at least in part, insufficiently regulated, as the organ retention scandal in the UK and Ireland revealed. When confronted with situations where the persons concerned cannot decide for themselves whether they want to contribute tissue or body material for research, we have to refer to notions of dignity or respect for persons. Although these concepts cannot always give us a perfect orientation on how to deal with the body material of deceased persons, at least it helps us to understand the concerns of relatives and the public when the interests of researchers, deceased persons, and their relatives are conflicting.

Principle of non-maleficence

In finding an appropriate regulation for tissue research, the principle of non-maleficence is of particular importance. Given that tissue research is often carried out on excess tissue which is left over from therapeutic interventions and would be thrown away anyway, it is a common perception that, as a rule, tissue research involves no severe risk of physical harm for tissue donors. Even where tissue is donated for other reasons than therapeutic interventions (e.g. tissue donation for epidemiological biobanks) it does not usually entail any significant physical harm. However, there is good evidence that considerations of physical harm are far too narrow to capture the real risks that donors undergo when, for example, they participate in biobank research. Some patient groups who donate body material (e.g. HIV patients) most certainly have a strong interest that the information about their health status is kept confidential. In times where employers carry out blood tests on their potential employees before providing them with a work contract, and health insurance companies are keen on learning more about the genetic predispositions of their prospective insurant, data from genetic analysis that are stored in the archives of epidemiological and disease-related biobanks might pose a threat to the life plans of the respective employee or insurant. Taking this into account, it follows that the risk analysis and the relevant fields for the application of the non-maleficence principle have to be adapted adequately to the handling of human tissue.

Informed consent

The field of tissue research has a problematic tradition of using tissue for research without the explicit consent of probands or patients. This common practice is also due to the sometimes unclear property status of material which was extracted from the human body and the undefined value of human tissue. Even excised and excess tissue from a therapeutic intervention may have practical value for the respective patient, for example when new treatment options make it necessary to reassess the status of tumour cells or genes. Even if this will probably not occur, the explicit consent of the proband or patient may be necessary (see the preceding remarks on the risks of genetic tissue research). Therefore, from our point of view, tissue research is (at least as long as body material has not been anonymized) 'medical research involving human subjects, including research on identifiable human material and data' as described in Article 1 of the Declaration of Helsinki.

Consequently, research on identifiable human material should be justified by the donor's informed consent.[3] There is an ongoing discussion on how detailed and specific informed consent needs to be in the case of biobank research, and whether a broad or even blanket consent of the donor would be acceptable as well. Naturally, this problem cannot be solved in the framework of the present volume. However, two important insights should be mentioned: first, the problem is less serious in the case of disease-related biobanks, where the use of tissue is limited to research on one or a specified group of disease entities. Under these circumstances it is possible to formulate an adequate consent for this type of research. Secondly, given that there are high security standards for the storage and confidentiality of data, good governance rules for the organization of biobanks, adequate and steady scientific and ethical review, and an institutionalized form of donor and patient participation in biobank governance, it would be far more acceptable to work with a broad consent in the case of epidemiological biobanks. However, in this context the question arises as to whether the framework for biobank research should be defined in a more binding way—on a national or supranational (i.e. European) level.

Trust in scientific research and the public interest

Biobank research establishes mutual obligations between the donors and the research institution which is at best based on a societal phenomenon called *trust*. The English language provides for an inspiring ambiguity in this context: *Trust* is not only the name for this societal phenomenon, but similarly for a public institution, *the trust*, which can be trusted and may then serve as a guardian for precious goods. When patients donate tissue for research, they normally expect that this will be an academic, and therefore non-commercial, public endeavour. Obviously, when no financial profit is generated from tissue research, suspicions regarding the potential exploitation of tissue donors become pointless. Interestingly, many share the normative premise (as, for example, is articulated in the *Statement on Benefit-Sharing* of the Human Genome Organization's Ethics Committee) that the individual genome is part of a more general entity—the human genome—which cannot be classified as individual property, but must rather be seen as a *common heritage of humankind*. This view would also be supported by the widespread reluctance to patent human genes. Both examples show that genes are regarded as a kind of intellectual property and that there is a general dissent as to whether the privatization of this resource should be allowed or whether it should be treated as a public resource. In the latter case it can be concluded that biobanks should predominantly be used by non-commercial research institutions, perhaps governed by a public trust, to build up a public monopoly on the beneficial effects that the use of these tissue collections may have.

---

[3] See also UNESCO International Declaration on Human Genetic Data, Art. 8 (a): 'Prior, free, informed and express consent, without inducement by financial or other personal gain, should be obtained for the collection of human genetic data, human proteomic data or biological samples, whether through invasive or non-invasive procedures, and for their subsequent processing, use and storage, whether carried out by public or private institutions. Limitations on this principle of consent should only be prescribed for compelling reasons by domestic law consistent with the international law of human rights.'

Solidarity

The acceptance of epidemiological biobank and tissue research also depends on societal experiences and traditions. It can be observed that the long-standing and vivid national tradition of health registries and epidemiological research in the Scandinavian and Baltic countries and in the UK also leads to a significantly higher acceptance of tissue and biobank research in these countries. One of the factors influencing this acceptance is the notion of solidarity, specifically when it is traditionally associated with communal health care. As mentioned above, there are indeed existing risks of biobank research for patients (which could theoretically be strongly diminished by society).

In a framework of solidarity, healthy or diseased persons are ethically obliged to donate their tissue for research purposes when this is beneficial for a specific group of patients or society in general. However, the problem with solidarity at the present time is that the previous waves of market liberalization and privatization make it questionable whether and how much solidarity can still be reasonably demanded. Additionally, solidarity is a mutual ethical obligation, and therefore the kind of benefit that patients or tissue donors could or should obtain from tissue or biobank research needs to be clarified. Last but not least, the question arises as to how solidarity could be promoted by the regulation of human tissue research.

Part II: Legal approaches

Property in the human body and the non-commercialization principle

Article 21 of the European Bioethics Convention enshrines a broadly recognized principle of European policy-making: 'that the body and its parts, as such, should not give rise to financial gain'. This concept is often referred to as the 'non-commercialization principle' or the 'non-property principle'. However, a closer look reveals some interesting paradoxes regarding this issue. First, although the commodification of the living body is indeed excluded by all European legal frameworks and constitutions, body parts which are separated from the human body can still become a thing according to many European jurisdictions and therefore might be regarded as transferable property. In Germany, for instance, a part of an organ that is separated from the human body becomes a thing and, after its processing, even a medicinal product. However, if we talk about 'the body and its parts' as tradeable medicinal products, it is by no means clear that the body must not 'give rise to financial gain'. Secondly, despite the European policy's preference for the 'donation model' (e.g. in the context of organ donation or as a governing principle in the EU Tissue Directive), there are a considerable number of private players which focus on 'financial gain' *after* the rather altruistically motivated act of donation. Therefore one has to distinguish between the *commercialization of the human body* (which is indeed excluded) from the *economization and trading of human tissue for research* (which is usually accepted). By taking this distinction into account, it becomes clear that the so-called non-commercialization principle is not nearly as restrictive as it seems to be at a first glance.

Convergence of ethics and law?

The practical cooperation between ethicists and lawyers is sometimes characterized by incomprehension of the work of the other discipline. Ethicists tend to assess the legal approach as too rigid and inflexible, whilst lawyers tend to characterize ethical texts as cloudy and expansive. Some medical lawyers are also very critical of the work and even the existence of ethical review boards as a kind of ad hoc committee with unclear competencies. However, current developments, at least in the fields of medical ethics and medical law, show a different tendency as a kind of approximation of ethics and law in the practical field. This highly interesting observation could lead to a totally new normative situation, i.e. that there is a relatively fluent observance of new developments in the practice of medicine which then, via ethical and legal deliberation, leads to a comparably smooth political regulation. The approximation of the two disciplines can also promote an exchange or combination of elements and approaches which were previously assigned to the other discipline.

European conventions and directives and national law

Currently, research with human tissue is not regulated on a European level. Given that in future the regulation of health-care issues in the European Union should stay under national responsibility as well, the question occurs as to whether there will be any mechanism of harmonization of tissue research in the coming years. For instance, the EU Tissue Directive (Directive 2004/23/EC) is solely concerned with the therapeutic use of human tissue and cells and consequently does not deal with human tissue research. There are also examples on the national level that human tissue research remains unregulated. In Germany, the Gene Diagnostics Law (Gendiagnostikgesetz) of 2009 is only concerned with the diagnostic use of genetic analysis and leaves out scientific research with human tissue. However, there are other documents, such as the European Convention on Human Rights and Biomedicine by the Council of Europe, which also contains provision that are relevant for research with human tissue. For example, Article 22 of the Convention demands:

> When in the course of an intervention any part of a human body is removed, it may be stored and used for a purpose other than that for which it was removed, only if this is done in conformity with appropriate information and consent procedures.

This means that not only a huge part of the ethical and legal discussion about the 'secondary use' of human tissue, but also the question of whether and what kind of informed consent is necessary for the donation of tissue for research, has to be seen in the light of this article. By 2009 the Convention had already been ratified by 24 member states of the Council of Europe. The fact that the Convention includes almost all East European countries (Bosnia, Herzegovina, Bulgaria, Croatia, Czech Republic, Estonia, Hungary, Lithuania, Moldova, Slovakia, Slovenia, and the former Yugoslav Republic of Macedonia) can well have an influencing effect on the other member states which have not yet ratified it. In fact, in the latter countries there seems to be a tendency to transfer single articles of the Convention into national law. For example, one can find a provision similar to Article 17 of the Convention *(Protection of persons not able to consent to research) in the* Section 41 (*Besondere Voraussetzungen der klinischen Prüfung [Special preconditions for clinical testing]*) of the German Drug Law, although Germany is one of the Western European countries which has not ratified the Convention.

Other influential sources can be national regulations in neighbouring fields, for example in organ transplantation. In Belgium, one of the arguments for an opt-out model for tissue research[4] was the experience with the existing opt-out model for organ donation, although this analogy is criticised by some scholars in the field. It can be concluded that whilst there are a multitude of influences on national legislation and regulations in the field, this does not exclude the emergence

---

[4] *Loi relative à l'obtention et à l'utilisation de matériel corporel humain destiné à des applications médicales humaines ou à des fins de recherche scientifique [Law on the procurement and use of medical application in humans or for scientific research]* (19 December 2008). Available at: http://vlex.be/vid/corporel-humain-applications-humaines-50448263 (accessed 28 January 2010)]

of similar tendencies in European regions and country groups. These existing convergences might display a potential starting point for a future harmonization of tissue research in Europe.

Cross-border transfer of tissue and body material

One of the crucial points for the realization of research projects which rely on international cooperation is the transfer of human tissue across national or even European borders. This challenges the interests of tissue donors in several respects. First, it has to be asked whether the donor's consent also covers the transfer of their body material to another country. Secondly, it needs to be clarified whether the donor's identity can be either directly or indirectly disclosed by the receiver of their tissue. Thirdly, what is the level of data protection in the other country? Does it match the regulations that are in force in the donor's country? As regards the latter issue, the existence of the European Data Protection Directive (Directive 95/46/EC) also has some implications for tissue research. Therein it is stipulated that personal data from within the European Union may only be transferred to third countries provided that an 'adequate level of protection' exists (Chapter IV, *Transfer of personal data to third countries*). From the national point of view, it is reasonable to argue that all special provisions for donors and patients should also be obeyed by the receiver of this tissue in another European country. To this end, the tissue 'sender' would have to use so-called 'material transfer agreements' which rule the level of data protection at the receiving institution. For example, this mechanism was developed by a German research group for the cooperation of biobanks in several European countries.[5] The result is a kind of 'private' form of research harmonization which is work intensive, but nevertheless effective.
Additionally, one can observe a kind of de facto harmonization in international tissue research and biobanking which consists of the following elements:

- transfer of property in body material from the proband or patient to the research institution;
- informed consent of the tissue donor for research with his body material;
- pseudonymization ('coding') of personal data and body material at the biobank, which will give no identifiable data or material to other institutions.

Homogenization of the legal situation in the European Union?

Given the importance of tissue transfer for international research projects, there are also strong tendencies to build up a common framework for research with human tissue and biobanking on a supranational level, at least in the European Union. However, some researchers caution against the harmonization of human tissue research regulation as this might have some problematic implications. For example, when there is a diversity of regulations, national biobanks will have a diversity of data and material stored, which can be useful for yet unknown future research projects. Researchers in countries with more liberal research regulations might also fear that a cross-national framework will turn out more restrictive and thus might restrain the freedom of research that was granted to them by their former national framework. However, given the importance of international cooperation, the protection of the interests of donors and patients as well as the confidentiality of their data, the arguments for a harmonized ethical and legal framework seem to outweigh such concerns.

[5] **Goebel, J.W., Pickardt, T., Bedau, M.,** et al. (2010). Legal and ethical consequences of international biobanking from a national perspective: the German BMB-EUCoop project. *European Journal of Human Genetics*, **18**, 522–5.

Part III: Practices: disciplinary perspectives

Tissue research in clinical practice and pathology

The discipline which classically deals with tissue samples in medicine is the subject area of pathology. As a rule, all important tissue samples which need further analysis and assessment go to a hospital's pathology department, and, because of this important position in the clinical network, local pathologists are often seen as mighty figures by their fellow tissue researchers. This is also why the handling of tissue samples by pathologists has a great influence on the general handling of human tissue in hospitals and medical research projects. In a way, the specific ethical and legal problems in the field of human tissue research can only be understood with some knowledge of the previous practices in the field of pathology. In the past, for example, informed consent was not a big issue, partly because the purpose for tissue analysis was well defined and mainly therapeutic, and secondly because tissue examination by pathologists had only local consequences which were mostly limited to the patient concerned. However, both characteristics have changed with the advent of the current concept of genetic tissue research: firstly, the purpose of the genetic analysis is now rather broadly described (corresponding to the discussion on 'broad consent' in tissue research) and non-therapeutic. Secondly, the research is done not only locally, but also in an international context, and without adequate safeguards for the protection of confidentiality, sensitive patient data might be distributed via unclear research networks. Therefore the existing changes in research on human tissue are not only a remarkable chance but also an important challenge for the discipline of pathology. The future will show whether adequate answers in relation to the interests of patients *and* research can be found from the side of this discipline.

Quality standards and normative implications from the perspective of medical ethics

It is an interesting observation from the perspective of medical ethics that apparently technical documents, which are supposed to have only an administrative or organizational content, sometimes have clear normative implications. This is obviously the case with the European Tissue Directive (Directive 2004/23/EC). Although the goal of the Directive is (merely) to set 'standards of quality and safety for the donation, procurement, testing, processing, preservation, storage and distribution of human tissues and cells', provisions regarding the public acceptance of tissue and cell donation, the preference for the mode of voluntary and unpaid donation, altruism in tissue and cell donation, and respect for and dignity of deceased donors can also be found. Two further important principles of the Tissue Directives are the provisions for the traceability of human tissue and cells and the identifiability of the tissue donor. Although established for the purpose of the safety of donated tissue and cells, these principles also have some normative implications. A similar tendency can be identified in the case of tissue and cell donation for therapeutic use as in the case of research on human tissue and cells: the prior anonymous use is transferred into therapeutic or research applications with pseudonymized or 'coded' material. A link between the donated material and the donor, which is actively maintained by the institution that is responsible for the storage or transfer of tissue, also remains. While, in the past, the anonymous use of human tissue and cells was often used as an argument that it is not necessary to seek the informed consent of the 'source' of the prevailing tissue, the newly established quality standards have changed this situation. In addition, the existing link between the donor and the material can now also be used to give more information to the donors as to how their tissue was finally used.

Professional discourse and patient's view on tissue research

The ethical and legal discourse and regulations regarding research with human tissue are grounded in a kind of idealization of the existing situation in applied research in the field. As everybody who works in the field of medical ethics and law knows, there is a remarkable discrepancy between ethical and legal theory and medical practice. Researchers and medical practitioners often have their own problems and worries, and do not necessarily see medical ethics and law as a support in their struggle for a high quality of research. Similarly, there is a clear difference between the purpose of ethical and legal guidelines for the protection of the interests of patients and the question of what they actually achieve in practice. For example, informed consent may be a sufficient condition for the protection of patient autonomy from the point of view of medical ethics, but it is an open question as to whether it is indeed sufficient in a concrete practical situation in the field. Therefore accompanying studies from sociology are an indispensable instrument for replenishing the theoretical point of view. Two questions in this field are of special importance. First, how are professional guidelines implemented for the practical application and in different national frameworks? Secondly, how are ethical and legal requirements perceived by the patients themselves?

Christian Lenk
Nils Hoppe
Katharina Beier
Claudia Wiesemann
Göttingen and Hannover 2010

Contributors

Nabila Ahmed
REPAIR-lab,
Institute of Pathology,
European Institute of Excellence
on Tissue Engineering and
Regenerative Medicine, University Medical
Centre Mainz,
Germany

Y. Michael Barilan
Department of Medical Education,
Sackler Faculty of Medicine,
Tel Aviv University,
Israel

Katharina Beier
Department of Ethics and History of
Medicine,
Georg-August-University Göttingen,
Germany

Milena D. Bister
Department of Social Studies of Science,
University of Vienna,
Austria

Jasper Bovenberg
Legal Pathways Institute for Health and
Bio-law Aerdenhout,
Netherlands

Christoph Brochhausen
REPAIR-lab,
Institute of Pathology,
European Institute of Excellence on Tissue
Engineering and Regenerative Medicine,
University Medical Centre Mainz,
Germany

Virginie Commin
Centre National de la Recherche Scientifique
(CNRS) Paris,
France

Viviana Daloiso
Institute for Bioethics, School of Medicine
"A. Gemelli",
Catholic University of the Sacred Heart
Rome,
Italy

Kris Dierickx
Centre for Biomedical Ethics and Law,
Leuven,
Belgium

Bianka S. Dörr
Faculty of Law,
University of Augsburg,
Germany and University of Zurich,
Switzerland

Austen Garwood-Gowers
Nottingham Law School,
Nottingham Trent University, UK

Christine Hauskeller
Senior Research Fellow,
ESRC Centre for Genomics in Society,
University of Exeter, UK

Nils Hoppe
Centre for Ethics and Law in the
Life Sciences (CELLS),
Leibniz Universitaet Hannover,
Germany

Nadja K. Kanellopoulou
HeLEX Centre, Department of PublicHealth,
University of Oxford, UK

C. James Kirkpatrick
REPAIR-lab,
Institute of Pathology,
European Institute of Excellence on Tissue
Engineering and Regenerative Medicine,
University Medical Centre Mainz,
Germany

Christian Lenk
Department of Ethics and History of Medicine,
Georg-August-University Göttingen,
Germany

José Miola
School of Law,
University of Leicester, UK

Remigius N. Nwabueze
University of Southampton,
School of Law Highfield, UK

Paola Parente
Institute for Bioethics,
School of Medicine "A. Gemelli",
Catholic University of the Sacred Heart Rome,
Italy

Nicolas Roßricker
REPAIR-lab,
Institute of Pathology,
European Institute of Excellence on Tissue
Engineering and Regenerative Medicine,
University Medical Centre Mainz,
Germany

Joseph Selling
Department of Theological Ethics,
Leuven, Belgium

Antonio G. Spagnolo
Institute for Bioethics,
School of Medicine "A. Gemelli",
Catholic University of the Sacred
Heart Rome,
Italy

Leen Trommelmans
Centre for Biomedical Ethics and Law,
Leuven, Belgium

Johannes J.M. van Delden
UMC Utrecht,
Julius Center for Health Sciences and
Primary Care,
Department of Medical Ethics,
Netherlands

Rieke van der Graaf
UMC Utrecht,
Julius Center for Health Sciences
and Primary Care,
Department of Medical Ethics,
Netherlands

Susanne Weber
Department of Strategy and Development,
Ludwig-Maximilians University (LMU)
Munich,
Germany

Dana Wilson-Kovacs
Research Fellow,
ESRC Centre for Genomics in Society,
University of Exeter, UK

Abbreviations

BMA	British Medical Association
CAT	Committee for Advanced Therapies
CHMP	Committee for Medicinal Products for Human Use
CHRB	Convention on Human Rights and Biomedicine
ECHR	European Convention on Human Rights
ECtHR	European Court of Human Rights
GCP	good clinical practice
GMC	General Medical Council
GMP	good medical practice
GTAC	Gene Technology Advisory Committee
HFEA	Human Fertilization and Embryology Authority
HRA	Human Rights Act
HTA	Human Tissue Authority
HUGO	Human Genome Organisation
IRB	institutional review board
MHRA	Medicines and Healthcare Products Regulatory Agency
MRC	Medical Research Council
MTA	material transfer agreement
SAMW	Swiss Academy of the Medical Sciences
UDBHR	Universal Declaration on Bioethics and Human Rights
UDHGHR	Universal Declaration on the Human Genome and Human Rights

Part I

Key concepts of the ethical debate

Chapter 1

The biomedical uses of the body: lessons from the history of human rights and dignity

Y. Michael Barilan

1.1 Introduction: the story of the Irish giant

Charles Byrne, also known as the Irish Giant, earned his living by public exhibitions of his extraordinarily tall stature (254 cm). In 1783, when his days were numbered (he was only 22 years old) he gave a group of undertakers a hefty sum of money and ordered that his body be buried at sea so as to avoid grave-robbing by anatomists. However, the surgeon John Hunter outbid the giant, paying the undertakers £500, and thus was able to add the skeleton to his anatomical collection, where it is still on public display in London (Gould and Pyle 1896:330–1; Richardson 1987:57–8). Some contemporary bioethicists campaign for the burial of Mr Byrne's bones, but to date the Hunterian Museum in London has not complied with their requests Dreger 2004: 116).

This anecdote sheds light on the special immunity from legal and ethical censure enjoyed by anatomy—then and now. Regulation of anatomy in the EU and other countries is a developing novelty of the past decade (e.g. German Medical Council 2003). Generally speaking, not only are anatomists not obliged to document the consent of donors, but in many countries the appropriation of 'unclaimed' bodies, including those of unidentified persons, for anatomical purposes is legal. Moreover, once a specimen is in the possession of an institute of anatomy, it may be processed, displayed, and trafficked all over the world without being subjected to legal or ethical scrutiny.[1]

This situation is quite striking, since non-consensual non-trivial use of anyone's body or body parts is typically considered a grave violation of human rights and dignity. Indeed, the Council of Europe's Convention on Human Rights and Biomedicine (1999) regards the use of human organs and tissues as an issue of human rights and stipulates it upon informed consent.[2]

In this chapter I will argue that contemporary bioethics and biolaw are blighted by inconsistencies ranging beyond anatomy and tissue banking, that the treatment of dismembered body parts, tissues, and corpses does not fall within the ambit of human rights, and that genuine incorporation of the notion of human rights and human dignity into the regulation of the dead human body and body parts entails changes of a kind that many people are not ready to accept.

---

[1] The increasing awareness for the necessity for informed consent in pathology and anatomy is documented in Chapter 14.

[2] Article 22, http://conventions.coe.int/Treaty/EN/Treaties/html/164.htm (accessed 1 October 2009). The convention has to be ratified by the single states to obtain legal force (CL).

1.2 The paradoxical ethics of the human body

Perhaps the most troubling paradox regarding the biomedical uses of the dead is the quasi-universal taboo on non-consensual harvest of organs from the dead in order to save life by means of transplantation. Some countries, such as Belgium and Spain, have embraced 'opting out' schemes, which do not require explicit consent. However, the justification for 'opting out' programmes is presumed consent. They do not allow for harvesting organs against the explicit refusal of the patient, even when this is the only means of saving the life of a needy person. In actuality, doctors do not harvest organs against the refusal of the family either. Organ transplantation is the only context in which physicians systematically refrain from saving their patients' lives only because of the refusal of people who have no legal power to either grant or withhold consent. One wonders why the Irish Giant can be exhibited for pleasure against his explicit wishes, whereas in the UK kidneys for live-saving transplantation are not harvested non-consensually from the dead.

An even stranger anomaly exists within this already odd set of practices related to the dead human body. Many laws on advanced directives do not cover pregnant women. Daniel Sperling, who exposed this exclusion, criticized it from the perspective of women's rights (Sperling 2008:Chapter 3). Readers who think that this is one more debate between 'pro-life' and 'pro-choice' world views are asked to contemplate the following hypothetical scenarios.

- In the first scenario, a pregnant woman leaves behind an advance directive which says that in the case of brain death, she wishes to be maintained on artificial life support only if the coming baby is male.
- In the second scenario, or thought experiment, a pregnant woman leaves behind an advance directive which says that in the case of brain death, her organs be donated for transplantation and save a few people, even at the expense of the life of her fetus.

In my personal experience, even those who subscribe to women's right of self-determination, hesitate regarding the first scenario and balk at the suggestion that the fetus be dispensed with in order to save lives by means of organ transplantation.

It may be inferred that the debate on abortion is not reducible to the question of the person status of the unborn. If this was the case, all those who do not consider it a person must allow—actually require—the sacrifice of the non-person in order to save lives. The case also demonstrates that the problem of abortion and other maternal–fetal issues are not reducible to the value of autonomy either. If this was the case, the woman's right to choose would be similarly upheld in all three cases—ordinary living wills, the condition of a male child, and the choice of donation at the expense of the unborn. Moreover, if the issue was reducible to the value of human life, not only the living wills would not be respected, but also a person's explicit refusal to donate organs for transplantation posthumously.

Conceptualizing the body as property is not likely to yield less contradictory results (see Chapter 9 for different legal paradigms of human body parts). The current discourse on 'the body as property' elicits four relevant observations. The first is that bioethics and biolaw sometimes treat the human body as the property of the person, and sometimes not—it is never 'property' in the manner that a lifeless chattel is (Waldron 1989:39; Hyde 1997; Dickenson 2007; Nwabueze 2007). The second is that the property-like status of the live body is substantially different from the property-like status of the dead body (Sperling 2008:Chapter 3). The third observation is that, especially in liberal societies, the freedom of choice regarding the conduct of a person's live body is broader than the authority he/she exercises over the uses of the dead body and amputated body parts. Testaments regarding the disposal of the dead body 'are not compelling [legally] unless voluntarily enforced by survivors' (ibid.:145). Finally, the limits set on the 'bodily use freedom principle' are derived from the value of 'human dignity' (Harris 1996:65). Utilitarians who conceptualize the body as property may arrive at the conclusion that, because saving life is more important than protection of property rights, the healthy have a moral duty to give organs (e.g. a kidney) in order to save the lives of the needy (Fabre 2006). Such radical ideas might gain further support from contractual (rather than natural) theories of property (e.g. Murphy and Nagel 2002).

Perhaps, instead of perceiving the body as property, it would be better to focus on the person as a proprietor—of self, body, and one's labour and creativity. Indeed, 'liberals' tend to derive the ethics of the body from a concept of self-ownership. Liberals describe any attempt to restrict personal freedom with regard to one's own body as if society dispossesses the person from absolute self-ownership (Sandel 2009:Chapter 3).

In addition to the very similar paradigms of body as property and self-ownership, one can approach the ethics of the body from the perspective of contract theory. Within this framework, the subject of an action (e.g. the body) is only an indirect warning that genuine and free consent is lacking. Thus, only choices that are free from duress and unconscionability deserve respect and protection (Fried 1981:Chapter 7).

While echoes from these paradigms are found in the discussion that follows, this chapter will look at the ethics of the human body from the perspective of human dignity and rights. The theory of dignity and rights that will be presented may offer itself as a more consistent and useful paradigm than 'the body as property', 'self-ownership', and contact theory.

1.3 **Human rights: basic interests and special restrictions on personal autonomy**

For a proper understanding of the notion of 'right' we have to distinguish between the formal structure of a 'right' and the idea of 'human rights'.

Following the legal scholar Hohfeld, it is now commonplace in philosophy and legal theory that 'rights' must conform to a certain logical structure (Wenar 2005). For example, according to a hypothetical contract between a tenant and a landlord, the tenant has the right to live in the apartment he/she rents and the landlord has a right to visit and inspect his/her property. These and other claims and duties of the tenant and the landlord embody the logical structure of rights. It is the very same logical structure which constitutes 'the right to life' and other human rights.

However, life is a universal and fundamental human good, whereas the right to the use of a specific apartment is not. The right of use is wholly derived from the contract and from the laws on contracts and on the uses of apartments. The right to life is not derived from any formal agreement or law. It is never hypothetical. Rather, whereas laws and contracts create rights such as in the contract and laws on rent, laws on the right to life presume that it already exists. Therefore rights like the right to life are referred to as 'natural rights'. In order to save life, one may suspend the rights derived from the rental contract (e.g. trespass in order to save people from a nearby fire), whereas the rental contract cannot authorize the violation of the right to life. Violation of ordinary rights may or may not embody a moral evil; violation of human rights is always considered a serious evil beyond 'public reason' and even beyond tolerance.

The 'right to life', as well as other human rights, is of a totally different character. It can be said that human rights comprise four levels: an *ethos*, an *internal formal structure*, an *external formal structure*, and a *substance*.

According to the ethos of human rights, all human beings share very similar vulnerabilities and needs, and each and every human being is endowed with unique dignity. The internal formal structure of human rights is identical to the logical format of all rights (Hohfeld's paradigm). The external formal structure of human rights creates a special relationship between rights and values and interests beyond the domain of rights. Whenever rights-related interests and values collide with non-rights-related interests and values, the former trump the latter (Dworkin 1984). Hence, human dignity and rights mark persons as 'inviolable'. The substance of human rights is universal and basic human needs and values.

We can conclude that natural human rights are construed as special moral (normative) properties of persons which protect and promote their very basic interests as humans, such as bodily integrity and freedom of expression.[3]

In addition to having a basic human interest at their centre, a unique feature of natural human rights is the central role of free choice (when the person is capable of choosing). The person has the power to decide what to say, what to do with their property, and how to care for their body and life.

Controversies hover over situations in which interest and free will collide with each other. Authors who emphasize the primacy of the will (i.e. regarding rights as essentially protecting free choice in certain domains, such as life) accept the notion of a right to suicide, to self-mutilate, or to perform in a freak show. Other authors reject the very possibility of protecting (let alone sustaining) choices that go against fundamental interests such as life and bodily integrity. This is the 'interest' doctrine of rights.

In politics, 'liberals' follow the 'will' paradigm, viewing human rights as liberties that people exercise in specific domains, whereas 'conservatives' follow an 'interests' pattern of thinking, granting 'human dignity' the power to exclude from the protection of rights personal choices that go against either basic human interests and values, especially life and the integrity of the person, or the very source of human rights—the notion of human dignity (Beyleveld and Brownsword 2001:25–7).

Elsewhere I have argued that human rights confer a very special kind of moral protection only when the relevant interest and the will go hand in hand (Barilan and Brusa 2008).[4] Therefore although much can be said in defence of the morality of assisted suicide for dying patients, a wish for euthanasia cannot be protected as a human right, nor can euthanasia of willing patients may

[3] The 'basic interest' condition in the theory of natural or moral rights differentiates it from interest theories of legal or positive rights, thus avoiding the problems associated with 'third parties beneficiaries' and rights within the criminal law which blight the latter. The power of the will in the basic interest paradigm does not necessitate the inclusion of powers to enforce compliance and to receive damages following violation.

[4] Sreenivasan (2005:267) and other theorists of rights refer to the 'advancement of a person's interest on balance', implying a monistic conceptualization of human goods according to which a basic interest can be forfeited for the sake of the person's overall good. For example, we may think of selling a kidney as forfeiture of the interest of health and bodily integrity in order to advance overall personal well-being. There are three key problems with this approach. First, it is not clear at all that a monistic conceptualization of human goods is tenable. Secondly, if selling a kidney is for the good of Simon's interest 'at balance', it should not matter whether he sells it for the purpose of transplantation, or for teaching medical students on a live kidney, or as food for animals. However, it is evident that the latter options are prohibited morally, even if Simon cannot sell his kidney for transplantation and cannot come up with an equivalent means of advancing his 'best interests at balance'. Thirdly, it is commonplace that the person knows best how to reckon his/her balance of interests; thus respect for personal will indirectly embodies respect for one's 'best interest at balance'. However, observance of human habits such as smoking, drinking, and reckless sex does not give empirical support to the 'person knows best' hypothesis.

be considered a violation of human rights. Wilful euthanasia might be right or wrong—it is not a matter of human rights. It can claim neither special protection nor strong condemnation. It falls within the ambit of 'public reason' (Rawls 1999:131–80; Barilan and Brusa 2010).

There is ample literature arguing the pros and cons of the morality of wilful alienability of rights-related interests. Little is said about certain kinds of self-alienation which nobody (or almost nobody) endorses, and that even liberal legislations and activists do not sanction. This *de facto consensus* among the contending camps ('liberals' and 'conservatives') is embodied by a few restrictions on self-alienation of rights-related interests. Conservatives support these restrictions because they do not endorse any kind of self-alienation of rights-related interests; liberals subscribe to these restrictions for reasons I will explore in the next section. These restrictions and the reasoning behind them are of utmost value for understanding human rights in the context of bioethics and biolaw.

1.3.1 **First restriction on self-alienation of rights-related interests: test of the endurance of the will**

Before patients are allowed to choose assisted suicide, donation or selling of organs, and similar acts that undermine interests protected by human rights, they are subjected to thorough evaluations and extensive counselling. Similar methods for verification of genuine choice exist in property law (a grace period allowing retraction from a contract, authentication by a notary, etc.). Here I focus on alienation of the right to bodily and sexual integrity of the person. Obligatory psychological evaluations prior to certain kinds of plastic surgery are one instance of tests for the endurance of the will.

Perhaps the Bible has such a law of authentication of the will for a person who chooses slavery (Exod. 21:5–6). The roots of this restriction in Western law are traceable to the Middle Ages, when prostitution was legal and tolerated as a 'minor evil'. Women under heavy financial duress were not allowed to enter into brothels (Perry 1985:149). Certain cities required periodic preaching to prostitutes by friars. These sessions were followed by assistance to restart a new life for those who chose to recant and renewed license for those who expressed the wish to stay in the brothel (Rossiaud 1988:Chapter 6). When people ply ordinary trades, society does not subject them to similar scrutiny. However, prostitution raises two concerns—that the person operates under duress, and that the person harms and degrades herself substantially. In order to tolerate such choices, one first has to authenticate them and, possibly, authenticate their autonomic depth. Both factors (authenticity of content and of depth) overlap significantly and bear directly on the other restrictions.

1.3.2 **Second restriction on self-alienation of rights-related interests: the real-time constraint**

I will call the second limitation the 'real-time' constraint. It requires the presence of consent at the time when a rights-related interest is being harmed. A typical example is the right of every participant in human experimentation to withdraw from the study at any time and with impunity, the damage to research notwithstanding. Whereas every person may undertake a commitment to hand over property in the future (e.g. let an apartment for rent), pledges on the body have no binding power. Change of the will does not revoke financial contracts. Rather, these very contracts exist so as to bind the participants in case they wish to back off. However, society does not recognize contracts on the body, such as slavery or prostitution (Hodson 1981). Even laudable acts such as altruistic donation of organs cannot be contracted, no matter how thorough and convincing the 'tests of the endurance of the will might be'.

The immediate implication of the real-time constraint is the impossibility of undertaking obligations to alienate rights-related interests.[5]

The real-time constraint has penetrated private law only recently, yet extensively. According to the *Hale Doctrine* a husband cannot commit rape against his wife since at the time of marriage she has given him the power over her body (Boyle 1981; Russell 1990:Chapter 2; McGregor 2005:Chapter 2; Martin et al. 2007; see also 1 Cor. 7:4). The legal right to divorce also expresses the impossibility of making autonomous choices whose very nature self-negates similar choices in the future. The world-wide criminalization of marital rape began in the 1970s, in parallel with the introduction of no-fault divorce and the relaxation of the laws on abortion (Glendon 1987).

Even wilful engagement in sex and desire for pregnancy do not create an irrevocable and unconditional commitment to carry the child to term. Pro-life scholars and activists do not gainsay the real-time constraint, but argue that it cannot surpass the very special value of human life. With the exception of the official teaching of the Catholic Church, humanity's religions and doctrines of ethics permit life-saving abortions (Barilan 2010b). This situation reflects a quasi-world-wide consensus regarding the impossibility of commitments to self-sacrifice even for the loftiest goals imaginable.

Common to all these changes is the recognition of the special value of the body, especially the sexual body, as a holistic locus of dignity, vulnerability, and non-violability. In contrast with private property, a pledge at one time cannot bind the person in the future.

The real-time constraint takes us beyond the debate on self-ownership. Although the person does not 'own' the body to the extent of having the power to alienate it in the future, the only agent that will exercise more power over the body in the future is the person him/herself—not society or another person.

1.3.3 Third restriction on self-alienation of rights-related interests: inseparability of the interest from the will

The next constraint is the inseparability of the will from the interest. Coercion of a person in his/her own benefit is generally unacceptable. This leads to care for suicide patients and force-feeding in critical starvation. However, tolerance of self-alienation of basic interests increases when the personal will invokes a basic human interest. Patients who ask for assisted suicide in the name of 'death with dignity' or those who go on hunger strike for the sake of freedom receive much sympathy. Opponents to assisted suicide and to hunger strikes will not permit such actions, but they are likely to respect the people involved. On the other hand, people who seek suicide for romantic reasons are treated as either deranged or morally reckless. Death with dignity and freedom are basic human goods (whatever their specific social construction might be) of the magnitude of shortening the life of terminal patients and of self-sacrifice in a just war.

1.3.4 Fourth restriction on self-alienation of rights-related interests: lack of expectations of self-alienation

The last constraint in our list is lack of expectation of self-alienation of rights-related interests. Society does expect people to sell property and to undertake unpleasant employments rather than becoming dependent on public charities for support; society does not expect people to prostitute themselves or to sell organs as a precondition to entitlement to charity. Even societies that tolerate

[5] Hegel argues that permissions such as the 'defence from necessity' are acceptable only because of a necessity of the immediate present. This is another aspect of the 'real-time' constraint (Hegel 2002, p.101 [1821: section 127]).

prostitution, surrogacy, and the selling of eggs and kidneys are careful not to force people to do so, even for the sake of child support and similarly important goals.

1.4 **Anatomy and tissue banking cannot be issues of human rights**

In this section I explore the use of dismembered body parts and dead bodies in the light of the above exposition of human rights. It will be argued that unethical use of dismembered body parts and dead bodies cannot violate human rights.

Since the human person is always embodied within a single spatiotemporal continuity, a simplistic scheme renders a disconnected body part a chattel whose natural owner is the person from whom the body part was taken.[6] Therefore it is his property or possession or trust, but it is not *him* (Penner 2000:Chapter 5). This attitude is borne out, for example, in Dickens' novel *Our Mutual Friend* (1864–1865), in which a character, Mr Wegg, who had lost a leg in a hospital amputation, sells the bones (which incidentally contain a rare anatomical variation) to an anatomy collector and later buys them back from him.[7] Following the amputation, the relationship between Mr Wegg and his leg is contingent; the ownership, or possession of, or the exercise of legal rights to certain interests manifested in the bone can be altered without invasion of Wegg's body or other interferences with his rights-related interests. The bone is physically—although, perhaps, not symbolically—alienated from Wegg. From the perspective of human rights, we might say that personal free will exists, but the interests in question are substantially different. The pecuniary, aesthetic, and sentimental values and uses of the dead and detached leg bones are quite remote from the uses and meaning of the living leg. Moreover, the special moral appeal of human rights is derived from the very close association between personal free will and basic human interests. Human rights protect the 'impersonated' body. Loss of this 'impersonation' substantially alters the interest and places it below the threshold of the power of human rights. Anything harmful or degrading or torturing that may be inflicted on Wegg's bones is not in the same metaphysical and moral categories as actions performed on his unamputated leg. Therefore the real-time constraint is something beyond a condition of temporality; it is also rooted in the inherence of personal free will in the impersonated body and the embodied person and in the unique role of the phenomenology of the present moment in the construction of the subjective self.

A human person is always a living individual extension in space. The person is embodied; the body is impersonated. The one-to-one relationship of embodiment is widely accepted—a person is never embodied in more than one body. On the other hand, some bodies may not be considered 'impersonated' (as some ethicists regard embryos and irreversibly comatose people), whereas others may be impersonated in a fragmentary manner (e.g. multiple personality disorder).

Since the person is embodied in only one spatiotemporal extension, we regard the whole of that spatiotemporal extension as the locus of impersonation (the person supervenes in the whole body, not only in parts thereof). Whatever is done to any part of this spatiotemporal extension is done to one person only. This is neither a biological nor a metaphysical assertion, but a pragmatic

---

[6] When a laboratory processes human tissue (e.g. making a cell line), it may also claim certain measures of property rights in the final 'product'. The balance of claims is beyond the subject of this chapter. At times it is not clear who the natural owner is, for example, of umbilical cord blood (Dickenson 2007, pp.94–102). Genetic data have so many potential claimants to 'ownership' as to render the whole concept of 'most reasonable candidate' void (Barilan 2010c, 2012).

[7] One may regard this episode as a pastiche on Hegelian 'objectification' (*Entäußerung*) through 'externalization' (self-alienation) and sublation (*Aufhebung*) ('reabsorption') (Miller 1987:12–30).

stance justified by human vulnerability and human agency. Anything done to one part of the 'embodiment bundle' is very likely to affect other parts substantially. The intricate physiological, immunological, and anatomical interconnectedness among all parts of the organism makes it impossible practically to separate out parts of the body which are essential or very significant for the person from parts that are not. As we all know, as dispensable as a tooth might appear, as long as it is part of the lived body, it may be a source of horrific pain and mortal illness.

Extreme pain and similarly substantial events such as illness, recovery, and amputation may alter the personality. Change over time is a fundamental feature of the phenomenology of the self. Dementia or mental illness may affect agency, transferring the person from a competent and accountable agent into an incompetent agent. According to the conventions of human agency, although a person may feel him/herself 'different', he/she may still claim debts owed to the 'former' self; society will regard the current 'self' accountable for deeds and commitments of the 'previous' self.

With the possible exception of extremely rare situations such as conjoined twins or 'brain transplantation surgery', for every live human body, there is only one 'most reasonable' relevant person (Barilan 2012).

Since the dead body and amputated body parts are not impersonated, the authority claimed over them by the person and the surviving kin is weaker relative to the power of the will over its own body and life. It is weak at least because disrespect for personal wishes pertaining to the post-mortem body or the severed body part do not entail coercion of or interference with the person.[8]

The notion that posthumous (or post-amputation) autonomous wishes deserve less respect than personal autonomy of embodied persons is compatible with contemporary court rulings. Execution of testaments is usually stipulated as being 'within the limits of reason and decency as related to the accepted customs of mankind' (Sperling 2008:173–83). Such limitations do not exist with regard to the private conduct of living people. Tolerance of certain personal choices may not reflect respect for autonomy, but only profound aversion to the suffering and humiliation associated with coercion and manipulation of persons (Barilan 2010a). It follows that although the law will not stop a person from conscientiously jumping in front of lions, legal systems do not uphold requests to feed a human cadaver to the beasts.

In sum, three fateful steps separate the dead body and the body part from the person and his/her human rights. The first is the 'depersonation' of the dead body and body parts—it is not part of the embodied person. Secondly, anything done to an amputated body part or a dead body cannot involve coercion or manipulation of persons. Thirdly, the dead body and amputated body parts are beyond benefit and harm. Since both the dead body and the dead person are beyond benefit and harm, the treatment of the dead cannot be protected by human rights (Barilan 2012).

Precisely because the treatment of tissues and cadavers cannot harm persons and cannot violate human rights, there is a strong case in favour of pushing aside autonomous choices regarding tissues and cadavers in favour of fundamental values such as human dignity (see Chapters 2 and 3)) and the common good.

---

[8] However, according to other authors, if personal wishes regarding the use of the dead body are not respected, people are used as mere means to an end, which is clearly an offence to their dignity (see Chapter 2).

1.5 Anatomy, tissue banking, and human dignity

The Universal Declaration of Human Rights states that human rights are derived from 'recognition in the inherent dignity ... of all members of the human family'.[9] The same idea appears in many other legal and academic documents. It belongs to what I referred to earlier as the 'ethos of human rights'. The precise meaning, or meanings, of human dignity in ethics and human rights theory are subjected to. However, all sources agree that the normative dimension of *human dignity* entails a special regard to human beings, differentiating them from animals and mere objects.

There is not a single known society which tolerates exploitation or mere trashing of human cadavers. One may find people who ceremonially eat the brains of their dead parents or feed them to vultures. However, each culture regards its own practice respectfully while reproaching alternative modes of disposition (e.g. cremation in lieu of burial). Therefore, although maltreatment of the dead cannot violate human rights, abuse of human remains is universally considered offensive to human dignity—the moral foundation of human rights. From this odd situation, we can draw a few lessons from the ethos of rights and dignity that are applicable to the ethics of anatomy.

In Dickens' novel *Our Mutual Friend*, Wegg's bones were in the possession of a dealer in anatomy and natural curiosities, Mr Venus. As a result, Dickens did not overstretch popular sensibilities regarding the use of human remains. Had Mr Venus given the bone to his dog to play with, Dickens' readers—nineteenth and twenty-first century alike—would revolt, for it is commonplace in Western ethics that human remains may only be used within scientific, religious, and educational contexts (in most other cultures, they may not be used at all).[10]

Indeed, the dissociation of the 'will' from the 'interest' offers dead and removed body parts one kind of protection that is stricter than that given to the live person. Whereas many ethicists believe that people may be allowed to pursue autonomous decisions to subject themselves and their bodies to contempt and abuse (having objects thrown at them for amusement and 'dwarf tossing', for example), dead and removed tissues may never be used in such a way. This is simply because the regulation of the use of human tissues and anatomical specimens does not entail coercion of an embodied free will. Policing Wegg's use of his intact leg is much more difficult ethically than regulating his conduct with his amputated bones.

1.6 The taboo on exploitation and the ethics of responsibility

Human rights and the idea of 'non-violability' bar the killing or harming of people, even in order to save many others. The notion that property rights can be violated in order to save life is highly contestable. But the dead human body and tissue are not chattel 'property'. Only after it is handed over for medical uses does the body and its parts undergo processes of objectification and commodification, some of which are morally and legally problematic (Steiner 2008). Outside the context of biomedicine, the body has no pecuniary value in Western law and ethics.

Since the dead body and dismembered body parts are not protected by human rights, there is a prima facie case for the utilization of the dead for the sake of issues related to human rights such as life-saving medical care.

At this stage it seems that my argument leads in the direction of non-consensual use of human remains for substantial biomedical purposes. However, in actuality, in the name of human rights

---

[9] See http://www.un.org/en/documents/udhr/(accessed 7 October 2009).

[10] Making art from human remains is controversial (Barilan 2005, 2006).

and the problematic 'right to privacy', there is a growing trend to condition any use of the dead body and tissue on informed consent.[11]

At a deeper level, this trend reflects a very strong sensitivity to exploitation, emphasizing it as the greatest evil possible in bioethics. Indeed, using people as 'mere means' has been defined as paradigmatically offensive to human dignity (see Chapters 2 and 3). Threat to life, bodily integrity, and sexual integrity (e.g. attempted rape) are still considered as just reasons for killing in self defence.

In my view, the concern with exploitation has increased out of proportion relative to other values and considerations. Since the treatment of cadavers and tissues is not of concern to human rights, any use of dead bodies and tissues cannot be exploitative in a morally significant manner. We must distinguish between the special powers a person exercises against exploitation, and the claim of authority over the future uses of removed tissues.

Every human being 'takes' his/her genes from his/her parents, and passes them on to his/her offspring. A genetic mutation or tissue typing (e.g. a specific HLA haplotype) is as relevant to the person from whose tissues a diagnosis has been made, as it is to every other person sharing the same genetic or biological make-up. Therefore a person's genes are not 'his' or 'hers', as he or she shares them with parents, siblings, and offspring.

Exposure of genetic and similar medical data might harm people through an assault on their vulnerabilities qua human beings. Exposure of medical or biological data *along with* the identity of a person is stigmatizing and is considered offensive to human dignity. Biological data should be differentiated from biographical information. The latter (which sometimes may be biological) informs about personal experience, ideas, choice, and behaviour—the most protected core of privacy. These are very serious concerns. However, we should keep in mind that a person's genes are not 'his' as much as his dismembered 'tissue' is not 'his' and his dead body is not 'his'. Of course, it is nobody else's either. It is part of a human heritage. Biomedical use of genetic and other biological material and data must be confidential and must serve scientifically sound ends in the benefit of humanity. In my view, the public exhibition of the Irish Giant does not count as a 'sound end in the benefit of humanity'. It is especially difficult to justify the exhibition of the authentic skeleton when a plastic model could be created in its stead.

Going back to the thought experiments presented at the opening of this chapter, one may recognize that pregnant women may object to their exploitation even for the sake of their own unborn children. However, when a woman tries to decide in such a way which child will live and which will die, her attitude becomes an exploitation of the power to thwart exploitation. There is a similar problem with the extension of people's legal powers over human remains, body parts, and genetic data beyond what is genuinely necessary for protecting their human rights and other salient interests. The person (and society) may restrict beneficial uses of human remains only in order to protect the person and especially his/her biographical privacy.

While attention to resistance of exploitation (even at the expense of saving life) has been taking place, the very concept of exploitation has been narrowing. According to many, free and informed consent purges an act from any hint of exploitation. This has led to the legalization of harvesting kidneys from live people (altruistic donors and sellers alike) without harvesting kidneys from cadavers non-consensually and to legalization of surrogacy without expansion, promotion, and encouragement of adoption.

[11] This trend can be observed in Austria, for example (see Chapter 17).

Over and above these practical considerations, if we accept a taboo on exploitation as the ultimate and inviolable foundation of bioethics, we are at serious risk of losing sight of biomedicine's promotive missions—life, health, and the care of the human body and person.

Tissue banking and anatomy involve significant moral risks, some of which may bring forth violation of human rights. Yet, we should also keep in mind that the very use of the dead, of human tissues, and of genetic material for the sake of the health of the living cannot violate human rights; all the while it promotes rights-related interests—life and health. Our first and foremost responsibility is for the well-being and dignity of the living, especially the vulnerable and the needy.

References

Barilan, Y.M. (2005). The story of the body and the story of the person: towards an ethics of representing bodies and body-parts. *Medicine Healthcare and Philosophy*, **8**, 193–205.

Barilan, Y.M. (2006). Bodyworlds and the ethics of using human remains: a preliminary discussion. *Bioethics*, **20**, 233–47.

Barilan, Y.M. (2008). The concept of responsibility: three steps within its evolution in bioethics. *Cambridge Quarterly of Healthcare Ethics*, **17**, 114–23.

Barilan, Y.M. (2010a). Respect for personal autonomy, human dignity and the problems of self-directedness and botched autonomy. *Journal of Medicine and Philosophy*, **35**, in press.

Barilan, Y.M. (2010b). Abortion. In R. Chadwick, H. Ten Have, and E. Meslin (eds), *Healthcare Ethics in the Era of Globalization*, London: Sage.

Barilan, Y.M. (2010c). Informed consent: between waiver and excellence in responsible deliberation. *Medicine, Healthcare and Philosophy* **13**, 89–95.

Barilan, Y.M. (2012). Ethics of anatomy. In R. Chadwick and G. McGee (eds), *Encyclopedia of Applied Ethics*, forthcoming.

Barilan, Y.M and Brusa, M. (2008). Human rights and bioethics. *Journal of Medical Ethics*, **34**, 379–83.

Barilan, Y.M. and Brusa, M. (2010). Triangular reflective equilibrium: a consciences-based method for bioethical deliberation. *Bioethics* (forthcoming).

Barilan, Y.M. and Turoldo, F. (2008). The concept of responsibility: three stages in its evolution within bioethics. *Cambridge Quarterly Healthcare Ethics*, **17**, 114–23.

Beyleveld, D. and Brownsword, R. (2001). *Human dignity in bioethics and biolaw*. Oxford: Oxford University Press.

Boyle, C. (1981). Married women—beyond the pale of the law of rape. *Windsor Yearbook of Access to Justice*, **1**, 192–213.

Dickenson, D. (2007). *Property in the body: feminist perspectives*. Cambridge: Cambridge University Press.

Dreger, A.D. (2004). *One of us: conjoined twins and the future of normal*. Cambridge, MA: Harvard University Press.

Dworkin, R. (1984). Rights as trumps. In J. Waldron (ed.), *Theories of rights*. Oxford: Oxford University Press.

Fabre, C. (1981). *Contract as promise: a theory of contractual obligation*. Cambridge, MA: Harvard University Press.

Fried, C. (1981). *Contract as promise: a theory of contractual obligation*. Cambridge, MA: Harvard University Press.

German Medical Council (2003). Recommendations on the treatment of human remains in collections, museums and public places. *DeutschesArzteblatt [German Medical Journal]*, August, C1532–6.

Glendon, M.A. (1987). *Abortion and divorce in Western law*. Cambridge, MA: Harvard University Press.

Gould, G.M. and Pyle, W.L. (1896). *Anomalies and curiosities of medicine*. New York: Saunders.

Harris, J.W. (1996). Who owns my body? *Oxford Journal of Legal Studies*, **16**, 55–64.

Hegel, G.W.F. (1821/2002). *Elements of the philosophy of right*. Newburyport, MA: Focus Publishing/ R. Pullins.

Hodson, J.D. (1981). Mill, paternalism and slavery. *Analysis*, **41**, 60–2.

Hyde, A. (1997). *Bodies of law*. Princeton, NJ: Princeton University Press.

McGregor, J. (2005). *Is it rape? On acquaintance rape and taking women's consent seriously*. Aldershot: Ashgate.

Martin, E.K., Taft, C.T. and Resick, P. A. (2007). A review of marital rape. *Aggression and Violent Behavior*, **12**, 329–47.

Miller, D. (1987). *Material culture and mass consumption*. Oxford: Blackwell.

Nwabueze, R.M. (2007). *Biotechnology and the challenge of property: property rights in dead bodies, body parts and genetic information*. Aldershot: Ashgate.

Murphy, L. and Nagel, T. (2002). *The myth of ownership: taxes and justice*. Oxford: Oxford University Press.

Penner, J.E. (2000). *The idea of property in law*. Oxford: Oxford University Press.

Perry, M.E. (1985). Deviant insiders: legalized prostitutes and a consciousness of women in early modern Seville. *Comparative Studies in Society and History*, **27**, 138–58.

Rawls, J. (1999). *The law of peoples*. Cambridge, MA: Harvard University Press.

Richardson, R. (1987). *Death, dissection and the destitute*. London: Routledge & Kegan Paul.

Rossiaud, J. (1988). *Medieval prostitution*. Oxford: Blackwell.

Russell, D.E.H. (1990). *Rape in marriage*. Bloomington, IN: Indiana University Press.

Sandel, M.J. (2009). *Justice: what's the right thing to do?* Cambridge, MA: Harvard University Press.

Sperling, D. (2008). *Posthumous interests: legal and ethical perspectives*. Cambridge: Cambridge University Press.

Sreenivasan, G. (2005). A hybrid theory of claim rights. *Oxford Journal of Legal Studies*, **25**, 257–74.

Steiner, P. (2008). Beyond the frontier of the skin: blood, organs, altruism and the market. *Socioeconomic Review*, **6**, 365–78.

Waldron, J. (1988). *The right to private property*. Oxford: Oxford University Press.

Wenar, L. (2005). The nature of rights. *Philosophy and Public Affairs*, **33**, 223–52.

Chapter 2

Exploring an alternative to informed consent in biobank research[1]

Rieke van der Graaf and Johannes J.M. van Delden

Biobank research is currently intensively debated. One of the ethical issues is to what extent the medical ethical requirement of informed consent can be met in this context. Obtaining genuine informed consent is problematic in biobank research for at least four reasons.

A characteristic of biobank research is that it uses human biological material that has been stored for a non-specific research purpose. At the time that human biological material is collected and stored the specific study in which this material will be used is generally unknown. Since human biological material is stored for future use, some years may elapse before participants can be informed about the research question that applies to their material. Therefore the first problem, which is widely recognized, is that participants in biobank research can only be given general information about the research for which their stored material will be used.

Secondly, informed consent with regard to the use of leftover material is problematic. Human biological material that is stored in biobanks can consist of either research material, which is material that is explicitly taken for medical research purposes, or material left over from diagnostic samples (e.g. from blood draws) or treatment samples (e.g. from surgery). Leftover material is not taken for research purposes, but only stored for this use. In other words, patients become research participants only if their leftover material is stored or used for research purposes. In the case of research material, it is self-evident that a researcher will ask informed consent before he/she can take material from a person. However, in the case of leftovers, the responsibility for informed consent may be unclear, since researchers are often not the persons who take the material from the patient. The physician who originally takes the material from the patient may not ask for informed consent because it is unclear in advance whether the patient's material will be stored. On the other hand, the researcher may not feel responsible because he/she uses material that would otherwise have been destroyed.

Thirdly, it has been unclear whether recontacting participants of biobank research to ask for informed consent contributes to respect for their autonomy. Some think that participants need to be recontacted at the time that their material is used since they cannot genuinely consent without project-specific information. Ethical guidelines often also require that a participant is recontacted if the material had originally been stored for another research project.

Fourthly, some think that the right to withdraw consent is problematic in biobank research. It is thought that if persons can withdraw, this may diminish the quality of research.

In this chapter we will study the extent to which informed consent research is appropriate in biobank. Informed consent is one of the fundamental requirements of research ethics

[1] In this chapter we have benefited from advice on biobank research from the Hospital Ethics Committee of the University Medical Center Utrecht in the Netherlands (2007).

(Emanuel et al. 2000; CIOMS 2002; WMA 2008). It is usually required to respect the autonomy of participants (Emanuel et al. 2000:2706; Beauchamp and Childress 2001:77). However, we will argue that informed consent is of limited value in biobank research (for a respect-based defence of a robust informed consent approach, see Chapter 3). We believe that the principle that people must not be used merely as a means is more useful for respecting the autonomy of participants in biobank research.

2.1 **Definition of biobanks**

We define a biobank as a collection of human biological material that is stored for medical research purposes. *Human biological material* is a collective term for blood, urine and other body fluids, tissue, and organs. As already explained, human biological material that is stored in biobanks can be divided into two main categories: research material and leftover material. Research material and leftover material can each be subdivided into new material that will be obtained in the future and archived material that has been collected in the past.

In biobanks, human biological material is stored for *medical research purposes*. Although this goal may seem obvious, it is a normative goal that distinguishes biobank material from other stored human biological material. Human biological material stored in biobanks should be distinguished not only from human biological material that is stored for diagnostics and treatment of patients (e.g. sperm banks), but also from human biological material that is stored with the sole purpose of gaining money or for making biological weapons). We agree with UK Biobank (2003, Section 8) that the primary aim of storing human biological material is medical research purposes which may generate and disseminate new knowledge that may ultimately benefit future patients.

Biobanks become valuable sources of stored material when the stored material is linked to information about the donor, such as sociodemographic, medical, genetic, familial, and lifestyle information. Anonymized human biological material is almost useless in biobank research. Moreover, if the donor of human biological material cannot be recontacted, it is impossible to ask additional questions for follow-up research. Most research ethics guidelines prohibit the use of identifiable material without the consent of the participant (Council of Europe 2005:articles 22 and 23; FMWV 2002). Therefore material in biobanks is usually coded. Coded material can be re-identified, preferably by a limited number of persons (UK Biobank 2003), and is regarded as unidentifiable for the researchers working with the material.

2.2 **Informed consent and biobank research**

As regards informed consent, guidelines on human subject research usually require that (i) researchers provide participants with information about the aims, methods, risks, and benefits of a study, (ii) participants must freely consent on the basis of the information provided, and (iii) participants always have a right to withdraw their consent (Emanuel et al. 2000:2706–7; WMA 2008:para. 24). However, criteria for informed consent appear difficult to meet in biobank research. This difficulty is also reflected by the legal provisions on human tissue research in the UK and France (see Chapters 8 and 11).

2.2.1 **Unknown research purposes**

A characteristic of biobank research is that when material is collected from participants neither the participant nor the researcher knows for what specific purpose this material will be stored and used. Participants can only be informed about the reason for storing the material (i.e. medical research purposes), and not about the specific aims of a study.

One of the responses to this problem is questioning the relevance of ethical and legal guidelines on human subject research for the purposes of biobank research. Some claim that the guidelines are not applicable (Dutch Ministry of Health, Welfare and Sports 2007:22). Others question whether the current guidelines on human subject research are the best way to protect the autonomy of participants (Eriksson and Helgesson 2005). Eriksson and Helgesson believe that biobank research, unlike experimental research with human beings, involves no risk of direct physical harm. Therefore they claim that 'the right to decide for oneself is virtually indispensable' in the context of medical experiments on human beings, 'but seems out of place' in research with bodily samples (ibid.:1073). We will discuss below that the principle of not using people merely as a means provides an argument for their suggestion.

Another more common response is asking for *broad* instead of *informed* consent (Hansson et al. 2006; Wendler 2006). In the case of broad consent participants are asked consent on the basis of information about the main lines of future research with their material. We agree that broad consent is more appropriate than informed consent in a context where the specific research purposes for the stored material are unknown. Currently, however, it is considered as a practical solution to the problem of not being able to meet the criteria for informed consent. Moreover, broad consent may lead to researchers having complete freedom to do what they like with the collected material (Hunter 2006; Maschke 2006). In order to avoid this scenario, some propose that participants should consent to specified medical research fields for which their stored material may be used. On the other hand, this option may restrict the research possibilities with the material too much. Moreover, it remains a form of broad consent because the specific research projects that will be undertaken with the stored material will not be known. Therefore asking participants to consent to using their material in specific research fields may not be particularly helpful. Others propose a charitable trust model for broad consent (Winickoff and Winickoff 2003; Hansson 2005). In this model people consent to donate their material to the biobank which 'has a responsibility to serve as a trustee' (Winickoff and Winickoff 2003:1182). According to Winickoff and Winickoff the trustee is, amongst other things, responsible for coding identifiable information before it is sent to researchers and for the review of research using the material by institutional review boards (IRBs). Furthermore, the biobank modelled as a charitable trust should be governed by a board of trustees and allow participants as members of this board, inform donors through websites about the research projects of the biobank, and require community consultation in the case of donation of particular populations, for instance ethnic groups (ibid.:1182–3). A trust 'can accommodate and foster the altruism, good governance, and benefit to the public' (ibid:1183). We think that the charitable trust model has merit, and in the following we argue that this model can guarantee respect for autonomy.

2.2.2 **Leftover material**

Informed consent for storage and use of leftover material may be problematic. It may create a situation where no one feels responsible for informed consent. The physician who originally takes the material from the patient may not feel responsible since it is unclear in advance whether this material will be used. Researchers may not feel responsible since they use material that otherwise would have been destroyed.[2]

In this situation some may rely on regulations that permit a waiver of consent under certain conditions, such as those in guideline 4 of the CIOMS guidelines (CIOMS 2002). However, we

[2] However, a practical suggestion for the establishment of a formalized informed consent procedure is made in Chapter 14, section 14.5).

think that a waiver of informed consent is not a structural solution for biobank research. Biobank research has become a practice that increasingly incorporates leftover material. Therefore we need more than exceptions to a rule.

Another answer to the problem of informed consent with regard to leftovers is relying on presumed consent. Presumed consent means that every patient is a potential donor unless he/she objects or opts out. However, like research participants who broadly consent to storage of their new material, patients who give presumed consent to store leftovers can only opt out on the basis of general information on biobank research. Hence obtaining presumed consent for storage of leftovers may be as problematic as obtaining informed consent for new research material if the interests of donors of leftovers are not further protected.

In sum, we believe that both waiver of consent and presumed consent are suboptimal solutions for respecting the autonomy of patients in the case of leftovers. Moreover, both the researcher and the physician will benefit from a situation where the responsibility for informed consent with regard to leftovers is clear. We shall return to this point below.

2.2.3 Recontacting the participant

It is unclear whether and to what extent informed consent requires that the researcher should try to recontact the participant in order to obtain informed consent when his/her material is used in a research project. Data from empirical research show that participants are not interested in being recontacted (Hansson 2005; Wendler 2006). Wendler (2006) proposes a 'one-time general consent' for biobank research: 'data from more than 33,000 people around the world support offering individuals a simple choice of whether or not their samples can be used for research purposes, with the stipulation that an ethics committee will decide the studies for which their samples are used'. Hansson also thinks that donors may not be interested in being recontacted. He refers to a survey in Sweden where people 'were more concerned with the security of the procedures and that secrecy was maintained than in the kind of research for which their samples were used' (Hansson 2005:416).

Some might argue that the donor should be recontacted since informed consent cannot be genuinely obtained when the material is collected and stored. They think that it has to be obtained when the specific research goals are known. However, is the first consent weaker than the second?

Recontacting may also be necessary for another reason. Informed consent may be given in the case of specific research projects. It is not self-evident that the participant approves use of his/her material for another research project. According to many guidelines, the proposed use should be in line with the previous consent (CIOMS 2002:guideline 4; UK Biobank 2003:para. 2; Council of Europe 2005:paras 21 and 22.1(i) and (ii)). If the proposed use does not fall within the scope of the previous consent, reasonable efforts should be made to recontact the donor of the material. We think that recontacting the participant is reasonable when research material has been collected for a specific research project. However, in the case of research material that has been collected specifically for storage in biobanks, most research projects will fall within the scope of the previous consent since participants can only broadly consent to biobank research. When broad consent has been obtained, recontacting the participant will hardly be applicable in practice. Therefore it is questionable how requirements for recontacting participants in the case of new research material may help to respect the autonomy of participants.

2.2.4 Withdrawal of consent

Most guidelines on research ethics state that donors have a right to withdraw their consent (Council of Europe 1997:article 5; WMA 2008:para. 24; CIOMS 2002:guidelines 4 and 5). Some argue that withdrawal of consent may be problematic in biobank research. Withdrawal may not always be possible. When a participant withdraws consent the material is usually destroyed or anonymized. However, when material is being used in studies at the time that a participant withdraws consent, it cannot be fully destroyed, nor can there be any guarantee that it can be completely anonymized. Eriksson and Helgesson (2005:1074) warn that if only the code keys are destroyed it may still be possible to identify a participant when the material has been linked to his/her medical information. Some also fear that withdrawal of consent may lead to loss of research quality (Helgesson and Johnsson 2005:321). Eriksson and Helgesson (2005:1076) think that withdrawal from biobank research should not be taken lightly. They argue that participants are not 'free riders' who only receive benefit from research outcomes with the human biological material of others (ibid.:1075). They propose that participants have a right to withdraw consent if they 'can present sufficient reasons why it is no longer reasonable to ask for...continued participation...If sufficient reasons are lacking, continued research is permissible'. According to the authors participants give sufficient reasons if they express 'genuine, deeply felt concerns...not based on misconceptions' (ibid.).

We think that only a limited number of persons will withdraw. The empirical studies mentioned earlier (Hansson 2005; Wendler 2006) show that people are not particularly interested in their material after it has been taken provided that it is used properly. If withdrawal turns out to be problematic for researchers, it should be reconsidered. Furthermore, in the following we will discuss the moral reasons why participants should not withdraw from biobank research. Unlike Eriksson and Helgesson, we do not support arguments of free-riding and asking participants to express deeply felt concerns at the time that they would like to withdraw. Instead we will argue that the principle of not using people merely as a means provides a rationale for not withdrawing consent.

2.3 **Interim evaluation of informed consent and biobank research**

The examples in the previous section demonstrate that informed consent is of limited value in biobank research. Answers to the problem of informed consent in biobank research seem to focus too much on the question of the extent to which informed consent can be met. What has been overlooked is the question of how we can respect the autonomy of participants in biobank research.

We have argued elsewhere (van der Graaf and van Delden 2010) that in order to respect autonomy in the context of research ethics, the principle of not using people merely as a means, henceforth the 'not merely as a means principle' (NMMP), may be more helpful than the informed consent principle. The NMMP can be grounded in Immanuel Kant's (1724–1804) notion of dignity. In his *Groundwork of the Metaphysics of Morals* (Gregor and Wood 1996:435), Kant states that autonomy is the ground of the dignity of rational agents and that agents who have dignity cannot use themselves or others merely as a means.

2.4 **Interpretations of not using people merely as a means**

Most scholars who explain the Kantian principle (O'Neill 1985; Korsgaard 1996; Wood 1999; Kerstein 2009; Parfit, forthcoming; Scanlon, forthcoming) refer to Kant's *Groundwork of the Methaphysics of Morals* (Gregor and Wood 1996:429–30). In these pages, Kant claims that people are used merely as a means in the case of false promises. Interpreters of these pages think that an agent *cannot consent* to a false promise and hence *cannot share the end* of the agent who is making this false promise. In other words, in order to avoid using people merely as a means, at least the conditions of *possible consent* and *end-sharing* have to be met.

2.4.1 Possible consent

In order to understand the condition of possible consent it is helpful to consider Derek Parfit's *Lifeboat* example (Parfit, forthcoming:92).

> A single person, White, is stranded on one rock, and five people are stranded on another. Before rising tide drowns these people, I could use a lifeboat to save either White or the five. These people are all strangers to me, and they do not differ in any other morally relevant way.

If the choice were White's, Parfit argues, she 'would have sufficient reasons to consent to my failing to save her life so that, in the time available, I could save the five' (ibid.:95). He thinks that 'the consent principle does not condemn this act' and hence that White is not treated merely as a means (ibid.:121).

Thus it is morally problematic to use people merely as a means if they do not have *sufficient reasons* to consent. Both Parfit and Thomas Scanlon also think that the kind of consent that is implied in the NMMP differs from actual consent. Scanlon opposes the more general consent of the NMMP to particular cases where actual consent is needed because 'my plan depends on some active contribution from you' (Scanlon, forthcoming:23). According to Parfit the consent in the NMMP differs from actual consent in that the first is a form of irreversible consent, whereby persons restrict their future freedom and 'not later learn some fact that might give [them] a decisive reason to regret the earlier consent' (Parfit, forthcoming:99).

2.4.2 End-sharing

Samuel Kerstein thinks that an agent *cannot share* another agent's end when the act of pursuing this end prevents him/her from attaining other ends that he/she are pursuing (Kerstein 2009). He believes that one cannot will an end and at the same time will another end that thwarts one's former end. If a person is not willing to give up his/her end in order to share the end of the agent who is using him/her, the former person is used merely as a means.

2.5 Merely as a means and biobank research

In the previos section we have set out that the kind of consent implied in the NMMP has most relevance in a situation like *Lifeboat*. According to Parfit, in this situation we have to choose between many possible acts that would have significant effects on many people whose interests and aims conflict and where there is at least one possible act to which everyone has sufficient reasons to consent. Thus we have to discuss whether participating in biobank research can be compared to *Lifeboat*.

The situations are equivalent to a certain extent. First, there is more than one possible act for participants of biobank research: to participate, not to participate, or to participate and to withdraw later on. Secondly, these acts have significant effects on many people whose interests conflict, at least to a certain extent. Future generations may benefit, participants risk abuse of sensitive information which may lead to refusal by insurance companies or discrimination by employers (Eriksson and Helgesson 2005:1072), and the results of biobank research may lead to stigmatization of the group to which the participant belongs (ibid.; Maschke 2006:193). However, unlike *Lifeboat*, it is not self-evident in the case of biobank research that participation is one of the possible acts to which a participant has sufficient reasons to consent.

We suggest that there are sufficient reasons if all stakeholders of biobank research (participants, researchers, sponsors, and others) agree that both storage and distribution of human biological material in biobanks is ethically acceptable. In our view stored material only counts as a biobank if it is stored for medical research purposes or will be used for this purpose, as we set out earlier.

As regards the distribution of the material, we think that the stakeholders should agree that the material is distributed according to the basic requirements for ethical research. We think that, apart from the usual ethical requirements such as protecting privacy, scientific validity, and value of research (Emanuel et al. 2000), stakeholders should agree on requirements related to distributive justice. In the case of biobank research distributive justice seems particularly relevant since many people with different cultural, social, and ethnic backgrounds participate in biobank research, many researchers and physicians will contribute to the collection stored in biobanks, and many research projects are possible with the stored material. As regards distributive justice, stakeholders have to consider, amongst others, whether subjects are fairly selected (Emanuel et al. 2000; CIOMS 2002), whether proposed research projects address health needs and priorities (CIOMS 2002), and whether researchers who contribute to biobanks receive a fair share of the use of the material. Obviously, further research is needed to consider on which requirements the stakeholders should agree. We propose here only that the stakeholders agree on a general level which ethical principles are relevant and to what extent they can be met.

We suggest that the charitable trust model, as discussed earlier (Winickoff and Winickoff 2003), should be taken into account in this context. When modelled as charitable trusts, biobanks should be able to determine and guarantee whether sufficient reasons exist. We have set out earlier that the charitable trust model involves all stakeholders. We think that these stakeholders should evaluate whether storage and distribution is reasonable with regard to biobank research. Winickoff and Winickoff (2003) present the charitable trust as a model and do not indicate how it should be realized exactly. We think that, in principle, IRBs should be able to judge from a reasonableness perspective whether participants could consent. Assigning this task to IRBs also has the advantage of not having to model every biobank in a hospital, which can be a freezer with blood samples, into a charitable trust. On the other hand, we have discussed earlier that the task of a charitable trust is broader than ethical review of proposed studies with biobank material. Amongst others, a charitable trust should inform participants about proposed research projects and consult community representatives when research with biobank material is undertaken (ibid.:1182–3). These tasks may ask too much of IRBs. Therefore it is conceivable that this task should be assigned to other boards or committees in a clinical setting.

Our analysis of consent in the context of biobank research has several implications. First, there is a solution to the problems of meeting informed consent in biobank research. We have seen that consent in biobank research nearly always takes the form of broad consent. However, as long as sufficient reasons for all stakeholders exist, broad consent can be grounded. In that case broad consent is no longer a pale shadow of informed consent.

Secondly, the analysis has implications for recontacting the participant and for withdrawal of consent. Parfit has argued that the consent implied in the NMMP is a form of irreversible consent. Furthermore, Scanlon has argued that actual consent is needed when an active contribution is asked. In this respect actual consent differs from the consent implied in the NMMP. Active contribution is hardly needed in biobank research, whereas it is central to experimental research involving human beings (Eriksson and Johnsson 2005).

The consent implied in the NMMP is irreversible and not actual. Therefore recontacting the participant for every proposed research project with the stored material becomes redundant in the case of projects where broad consent is obtained. Only where specific informed consent is obtained, should the participant be recontacted if the proposed use does not fall within the scope of prior consent. Furthermore, since in principle a participant gives irreversible consent, withdrawal from research is unreasonable. We have discussed earlier that Eriksson and Helgesson (2005) propose that participants present sufficient reasons when they want to withdraw. This may be too demanding. We think that withdrawal can be considered as unreasonable since all stakeholders, including the participant, have agreed in advance that there were sufficient reasons to participate. If withdrawal turns out to be problematic, or even impossible, we cannot say that the participant is used merely as a means.

On the other hand, we do not expect withdrawal of consent to be a problem of major concern, as we set out earlier. Furthermore, we agree with Helgesson and Johnsson (2005:321) that a right to withdraw is necessary to maintain public trust in biobank research. Therefore we think that people should have a right to withdraw.

In order to avoid using participants merely as a means they should not only have sufficient reasons to consent, but also share the ends of the agents who use them for their purposes. We have set out earlier that a primary end of biobank research is storage of human biological material which may generate and disseminate new knowledge that may ultimately benefit future patients. Although this is an obvious goal, we should not simply take for granted that ends of biobank research are shared.

In the case of new research material it may be possible to inform participants of the ends of biobank research and to consider in an interview whether participants share these ends.

As regards archived research material and leftover material, ensuring end-sharing will be more difficult. Recontacting participants will not always be possible. In such cases we think that biobanks as charitable trusts should broadly provide information about the ends of research projects with biobank material on websites and through other media.

Ensuring end-sharing also needs attention in the case of new leftover material. It may be practically unfeasible and also overly demanding for biobanks to interview patients about the ends of biobank research at the moment that they are hospitalized (for an empirical study of patients' and researchers' perceptions of informed consent before surgery, see Chapter 17). Furthermore, we think that end-sharing sets limits on the value of presumed consent for biobank research. Not all patients will read information brochures or websites before they enter a hospital. The best way to ensure end-sharing in the case of new leftover material may be to ask for broad consent in an interview after researchers have decided to collect and store the leftover material for research purposes. Since researchers depend on the material that is taken by physicians, we think that researchers and physicians will benefit when the clinical setting in which they work regulates the responsibility for consent with regard to leftovers.

2.6 **Conclusion**

In this chapter we have studied the extent to which informed consent in biobank research is appropriate. We conclude that informed consent is of limited value for respecting the autonomy of participants in biobank research. In the context of biobank research consent nearly always takes the form of broad consent. We claim that broad consent is an alternative to informed consent if the principle of not using people merely as a means has been met. This principle requires that participants must have sufficient reasons to consent to biobank research and hence be able to share the ends of biobank researchers. Sufficient reasons form the ground for broad consent. Furthermore, the principle of not using people merely as a means requires that broad consent is more than the mere provision of general information on biobank research. It will also be necessary to ensure, preferably in an interview, that participants and researchers both share the ends of biobank research. We have argued that sufficient reasons and end-sharing could be guaranteed by means of a charitable trust model. Although participants in biobank research risk minimal bodily harm, sufficiently protecting autonomy in biobank research should not be lost from sight.

References

Beauchamp, T.L. and Childress, J.F. (2001). *Principles of biomedical ethics.* Oxford: Oxford University Press.

CIOMS (Council for International Organizations of Medical Sciences) (2002). *International ethical guidelines for biomedical research involving human subjects.* Geneva: CIOMS.

Council of Europe (1997). *Convention for the protection of human rights and dignity of the human being with regard to the application of biology and medicine: convention on human rights and biomedicine.* Available online at: http://conventions.coe.int/treaty/EN/Treaties/Html/164.htm (accessed 22 December 2009).

Council of Europe: Steering Committee on Bioethics (CDBI) (2005). *Recommendation on research on biological materials of human origin.* Strasbourg: Council of Europe.

Emanuel, E.J., Wendler, D., and Grady, C. (2000). What makes clinical research ethical? *Journal of the American Medical Association,* **283**, 2701–11.

Eriksson, S. and Helgesson, G. (2005). Potential harms, anonymization, and the right to withdraw consent to biobank research. *European Journal of Human Genetics,* **13**, 1071–6.

FMWV (Federation of Biomedical Scientific Societies) (2002). *Code for proper secondary use of human tissue in the Netherlands.* Rotterdam: FMWV.

Hansson, M.G. (2005). Building on relationships of trust in biobank research. *Journal of Medical Ethics,* **31**, 415–18.

Hansson, M.G., Dillner, J., Bartram, C.R., Carlson, J.A., and Helgesson, G. (2006). Should donors be allowed to give broad consent to future biobank research? *Lancet Oncology,* **7**, 266–9.

Helgesson, G. and Johnsson, L. (2005). The right to withdraw consent to research on biobank samples. *Medicine, Health Care and Philosophy,* **8**, 315–21.

Hunter, D. (2006). Autonomy and majority rules have been misunderstood. *British Medical Journal,* **332**, 665–6.

Kerstein, S. (2009). Treating others merely as a means. *Utilitas,* **21**, 163–80.

Korsgaard, C.M. (1996). *Creating the Kingdom of Ends.* Cambridge: Cambridge University Press.

Maschke, K.J. (2006). Alternative consent approaches for biobank research. *Lancet Oncology,* **7**, 193–4.

Ministry of Health, Welfare and Sports (2007). Beleidsbrief ethiek. The Hague 22. Available online at: http://www.rijksoverheid.nl/documenten-en-publicaties/kamerstukken/2007/09/10/beleidsbrief-ethiek.html (accessed 24 September 2010).

O'Neill, O. (1985). Between consenting adults. *Philosophy and Public Affairs,* **14**, 252–77.

Parfit, D. *Climbing the mountain* (forthcoming). Preview available at: http://individual.utoronto.ca/stafforini/parfit/parfit_-_climbing_the_mountain.pdf (accessed 22 December 2009).

Scanlon, T. *Means and ends* (forthcoming). Preview available at: http://philosophy.fas.nyu.edu/docs/IO/1881/scanlon.pdf (accessed 22 December 2009).

UK Biobank, Ethics and Governance Framework (EGF) (2003). *Summary. Setting standards. The UK Biobank Ethics and Governance Framework.* Manchester: UK Biobank.

van der Graaf, R. and van Delden, J.J.M. (2010). On using people merely as a means in clinical research. *Bioethics,* doi: 10.1111/j.1467-8519.2010.01820.x.

Wendler, D. (2006). One-time general consent for research in biological samples. *British Medical Journal,* **332**, 544–7.

Winickoff, D.E. and Winickoff, R.N. (2003). The charitable trust as a model for genomic biobanks. *New England Journal of Medicine,* **349**, 1180–4.

World Medical Association (2008). *Declaration of Helsinki. Ethical principles for medical research involving human subjects.* Available online at: http://www.wma.net/en/30publications/10policies/b3/index.html (accessed 22 December 2009).

Wood, A.W. (1999). *Kant's ethical thought.* Cambridge: Cambridge University Press.

Chapter 3

Respect as a precondition for use of human tissue for research purposes

Austen Garwood-Gowers

3.1 Introduction

In large parts of the world, medical orthodoxy has long been grounded on a classical paradigm of science that is largely mechanistic, atomistic, and reductionist rather than holistic in character. This orientation has contributed to medicine developing into a vast industrial complex, one of the characteristics of which has been to progressively develop and exploit ways of using the body to meet wider ends. This chapter will argue that such use is only acceptable where it is consistent with respect for the individual and will explore some of the key implications of this for discourse, practice, and governance concerned with the use of tissue for research purposes.

3.2 Normative standards

Through the World Medical Association, the international medical community responded to the medical atrocities carried out during the Second World War by creating such initiatives as the Declaration of Geneva Physicians Oath 1948, which requires medical professionals to swear that they will treat the health of their patients as their 'first consideration', and the International Code of Medical Ethics 1949, which stipulates that 'any act, or advice which could weaken physical or mental resistance of a human being may be used only in his interest'. The Association also built on the Nuremburg Code's protective emphasis in medical research by creating the Declaration of Helsinki 1964 (as amended most recently in Seoul 2008). Principle 6 of the Declaration states that: '[i]n medical research involving human subjects, the well-being of the individual research subject must take precedence over all other interests'. A number of other principles elaborate on some of the implications of this emphasis, including Principle 9 which stresses the need for special protection of the vulnerable.

The post-war period also saw important human rights developments at the inter-state level, such as the Universal Declaration of Human Rights 1948 and the European Convention on Human Rights (ECHR) 1950, both of which emphasize non-discrimination and provide bodily protection of the individual through their emphasis on freedom from inhuman and degrading treatment and on such rights as life, liberty, and private life.

More recently, inter-state initiatives specific to the protection of human rights in the health context have also been developed, not least in the wake of the capacity of new medical technologies to bring new threats to human life. The most significant of these instruments for immediate purposes are the Council of Europe's Convention on Human Rights and Biomedicine (CHRB) 1997, UNESCO's Universal Declaration on the Human Genome and Human Rights (UDHGHR) 1997, and the Universal Declaration on Bioethics and Human Rights (UDBHR) 2005. The core of all three documents includes a general emphasis on human rights and dignity, and a more specific emphasis on such values as integrity, non-discrimination, and equality. Thus, for example, Article 1 of the CHRB requires parties to 'protect the dignity and identity of all human beings and guarantee everyone, without discrimination, respect for their integrity and other rights and fundamental freedoms with regard to the application of biology and medicine', adding that '[e]ach Party shall take in its internal law the necessary measures to give effect to the provisions of this Convention'. Article 2 states that 'the interests and welfare of the human being shall prevail over the sole interest of society or science'. Article 26 may restrict these and many of the CHRB's other rights and protective provisions, but only in a manner prescribed by law and necessary in a democratic society for protection of one or more of public safety, prevention of crime, protection of public health, or protection of the rights and freedoms of others. Dute (2005:5) has suggested that the human dignity and identity aspects of Article 1 are so basic that restriction of them cannot be accepted. He has also concluded that Article 2 makes 'clear that the human being must never be used as a [mere] means to an end' (ibid.:8). Even if this were not the case, one may observe in the context of discussing both the CHRB and the ECHR that it is contrary to, rather than necessary in, a democratic society to dilute something as fundamental as bodily protection of the individual to meet the mere perceived needs of others.

The legal impact of international instruments at a national level varies with both the instrument and the jurisdiction concerned. International medical community provisions are of merely persuasive effect. However, they can clearly become more significant where used as a template to develop domestic legislation, as has often occurred, for example, in relation to the Declaration of Helsinki. The effect of inter-state initiatives is more varied. In some jurisdictions, such as France, international agreements take precedence over domestic provisions in the event of a conflict. In contrast, in the UK this is only so for EU law, although the ECHR has a very important status not only because the UK Parliament has bound itself, subject to powers of derogation, to act on the decisions of the European Court of Human Rights (ECtHR), but also because nearly all the rights protected by the ECHR form part of domestic law via the Human Rights Act (HRA) 1998 and ECtHR jurisprudence is utilized by the domestic courts to interpret these rights. There are no parallel methods of enforcing the CHRB in the UK. However, it must be borne in mind that CHRB provisions were designed partly as an aid to the development of ECtHR jurisprudence in cases involving the nature of human rights in medical contexts[1] and accordingly may be extremely significant in domestic interpretations of the HRA in medical cases. However, the point of focusing on international instruments is not merely that in some respects they tend to be legally significant but that they are also morally and politically significant. The aforementioned instruments specifically, when taken as a whole, generate a broad consensus of medical and state powers in favour of respect for the individual. One implication of this is that medical use of the body must either be genuinely required to protect that individual's overall interests or be justifiable by reference to the rights of others, rather than simply being an overt or covert attempt to meet the needs of others. A good example of the application to tissue use is the case of *S and Marper* v. *UK* [2009] 48 EHRR 50. Here the Grand Chamber of the ECtHR found that the UK had violated Articles 8 and 14 of the ECHR by allowing the police too much leeway in their retention of fingerprints and samples from persons who, having been arrested and charged, had subsequently been acquitted or had their charge discontinued.

[1] See, for example, *Glass* v. *UK* [2004] 1 FLR 101.

3.3 Defending a respect-based approach

Human rights instruments appear to view the individual as having an a priori right to be respected. In the light of a long history and future potential for abuse of power, they also appear to view mere means to an end treatment as politically and pragmatically unattractive. Nonetheless, in the medical context, as elsewhere, many authors have supported such use with utilitarian-based arguments (e.g. Harris 1970; Calabresi 1991) and justice-based arguments (e.g. Calabresi 1991; Harris 1997, 2002; Gunn *et. al.* 2000; Fabre 2006; Brownsword 2008). These arguments tend to be uncritical about ends that might be pursued and to ignore the negative effects that can stem from legally enforcing a conception of morality in the domain of something as fundamental and personal as the body (see further Garwood-Gowers 2008).

Arguments against a respect-based approach are also made on a class-specific basis. Attempts to justify lesser treatment of certain classes often focus on the embryo and fetus at one end and/ or the dead at the other. Sometimes they additionally or alternatively focus on those 'in-between classes' who have certain disabilities or disorders. Palazzani (2008:89) notes that, whilst many ethicists embrace 'personalism' within which the term person and human being are synonymous, others embrace 'personism', within which persons are treated as a narrower and preferred class—often on the basis that only they, rather than human beings as a whole, exhibit the capacity for rationality or, more narrowly, awareness. Personism is problematic not only because it is focused on what potentially divides us as human beings and as aspects of reality in general, but also because it generates difficulties of definition and measurement. In particular, it is hard to see how one can definitively measure lack of awareness and hence, for example, declare a patient to have so-called persistent vegetative state (McCullagh 2004).

3.4 Respect as a basis for a robust approach to information- and consent-related standards

Shortfalls of protection in governance and practice in this area particularly relate to the applicability and nature of information and consent standards in specific circumstances and in relation to specific classes. Samples are increasingly being obtained specifically for research purposes, not least in the context of local and national biobanks. However, a large proportion of them continue to end up being used for research purposes although they have been removed for other purposes. Price (2005) notes that this includes over 3 million samples removed annually for treatment or diagnostic purposes within the UK National Health Service alone. Article 22 of the CHRB states that:

> [w]hen in the course of an intervention any part of a human body is removed, it may be stored and used for a purpose other than that for which it was removed, only if this is done in conformity with appropriate information and consent procedures.

The basic need for Article 22 is elaborated in paragraph 135 of the Explanatory Report which states that:

> Parts of the human body are often removed in the course of interventions, for example surgery. The aim of this article is to ensure the protection of individuals with regard to parts of their body which are thus removed and then stored or used for a purpose different from that for which they have been removed. Such a provision is necessary in particular, because much information on the individual may be derived from any part of the body, however small (for example blood, hair, bone, skin, organ). Even when the sample is anonymous the analysis may yield information about identity.

Unfortunately, paragraph 137 of the Explanatory Report is rather muted in its assessment of the scope of Article 22, stating that:

> The information and consent arrangements may vary according to the circumstances, thus allowing for flexibility since the express consent of an individual to the use of parts of his body is not systematically needed. Thus, sometimes, it will not be possible, or very difficult, to find the persons concerned again in order to ask for their consent. In some cases, it will be sufficient for a patient or his or her representative, who have been duly informed (for instance, by means of leaflets handed to the persons concerned at the hospital), not to express their opposition. In other cases, depending on the nature of the use to which the removed parts are to be put, express and specific consent will be necessary, in particular where sensitive information is collected about identifiable individuals.

This paragraph treats limited provision of information as potentially adequate (i.e. a leaflet) and disingenuously associates non-opposition by patients or their representatives with consent. It also rather weakly proposes that the necessity for express and specific consent is selective without precisely identifying the kinds of circumstances in which it would or would not be required and why. In these regards it seems to promote the meeting of needs in a manner that is incompatible with respect for the individual. So, more blatantly, does Article 14 of the 2008 Additional Protocol to the CHRB concerning Genetic Testing for Health Purposes. It allows some scope for genetic tests to be performed on the previously removed biological materials of a person who cannot be contacted for the benefit of his/her family 'in accordance with the principle of proportionality, where the expected benefit cannot be otherwise obtained and where the test cannot be deferred'. Both these provisions should be read down given that respect for the individual is one of the pillars of the Convention, is specifically emphasized in the genetic testing context by the Additional Protocol (most notably Articles 1 and 3), and underpins *Rec (2006)4* of the Committee of Ministers to member states on research on biological materials of human beings.

Nonetheless, the idea of sacrificing respect to meet need is a common driver for perspectives on appropriate information provision and consent standards in the use of tissue for research purposes—sometimes explicitly so (see Chapter 14, section 14.4)—and one does not need to look hard to find domestic practices and laws that reflect its influence. Starting at the practice levels, the following three case studies are illustrative.

1. The NHANES III project: the last of a three-part series of medical and epidemiological studies carried out in the USA by the Centers for Disease Control and Prevention involving 40,000 people, of whom 29,314 had samples taken with a generic consent for storage and testing but without any more specific information or consent. By the mid-1990s the samples were being used for molecular research with DNA but the option of 're-consenting' participants at a cost of $2 million was rejected.
2. Genetic data collection and use under Iceland's Health Sector Database Act (No. 139/1998): The Act enabled genetic data stored by the National Health Service to be used for the database without regard for the wishes of its source and genetic material to be used on a presumed consent basis. It was abandoned after opposition, which included a successful legal challenge to its impact on the privacy of those with forebears on it (Icelandic Supreme Court judgment No.151/2003).
3. Practices of materials being routinely retained and used for research and other purposes in England and Wales following post-mortems, in particular material taken from dead children without parental consent or even knowledge.

Failures at Bristol Royal Infirmary and the Royal Liverpool Children's Hospital (Alder Hey) were respectively highlighted in *Learning from Bristol: the report of the public inquiry into children's heart surgery at the Bristol Royal Infirmary* (Bristol Royal Infirmary Inquiry 2001) and *The Royal Liverpool Children's Inquiry Report* (Department of Health 2001). However, it quickly became apparent that the problem was much more widespread. A census carried out by the Chief Medical Officer (Chief Medical Officer 2001) found that by the end of 1999, 54,300 organs and body parts from stillborn children and fetuses alone had been retained following post-mortem. The country-wide picture was explored in the Isaacs Report (HM Inspector of Anatomy 2003).

Many laws covering the use of tissue for research purposes allow consent to be broadly given rather than requiring it to be purpose specific (for a qualified vindication of broad consent, see Chapter 2). Some allow material removed for one purpose to be used for others without consent being given for this, and many dispense with the requirement altogether in certain circumstances (e.g. where a donor cannot be found). Many jurisdictions allow use of body materials for research and certain other purposes, such as transplantation, to occur post-mortem simply on the basis that, following reasonable enquiry, there was no evidence of the deceased having objected during his/ her life (strong opt out or strong presumed consent), or that additionally there was either consent of those with parental authority or next of kin (weak opt in) or no evidence following reasonable enquiry that they objected (weak opt out or weak presumed consent).

The law in England and Wales provides a good illustration of these problems. The Human Tissue Act 2004 (for a critical appraisal, see Chapter 8) requires appropriate consent for storage and use of the body for most purposes under section 1(1) in conjunction with schedule 1, part 1. This includes most research. However, storage for the purpose of research, and actual use of material for the purpose of research, in connection with disorders, or the functioning, of the human body is exempted from the requirement for 'appropriate consent' if the material has come from the body of a living person and is part of research that is ethically approved in accordance with regulations made by the Secretary of State, and is to be, or is, carried out in circumstances such that the person carrying it out is not in possession, and is not likely to come into possession, of information from which the person from whose body the material has come can be identified (sections 1(7)–1(9)). This provision represents a rejection of even the generic consent approach initially recommended by the Royal College of Pathologists in its *Transitional guidelines to facilitate changes in procedures for handling 'surplus' tissue and archival material from human biological samples* (Royal College of Pathologists 2001) The basis of this rejection appears to have been cost. Furness and Sullivan (2004) estimated that if one minute was spent on obtaining consent for each of the relevant tissue samples obtained by the NHS each year, an extra 1339 full-time staff would need to be employed. One might counter that an immensely disturbing state of affairs must have been reached for a principle as important as consent to be considered not worthy of spending a minute on per case.

The Act also contains several other exemptions from the section 1 requirement for appropriate consent. It is also true that sections 1(5) and 1(6) exempt bodies and materials where they are imported or of a person who died at least a hundred years before section 1 came into force. Section 47 gives a limited power, but not a duty, to those running the Armouries and various museums to return material that was less than a thousand years old when it came into force. Several museums have aboriginal ancestral body materials, amongst them the Natural History Museum which was initially reluctant to give those materials back but eventually did so using its section 47 power. Section 45 makes it possible to perform non-consensual analysis of DNA under certain conditions, whilst section 43(1) makes non-consensual post-mortem preservation possible to facilitate transplantation. In addition, where appropriate consent does apply, satisfying it typically entails mere adherence to a weak rather than a strong opt-in system, although the latter is adopted for public display and, barring excepted material, anatomical examination (see sections 2(4) and 2(5) for children and sections 3(3) and 3(4) for adults).

The overall picture is hard to reconcile with the protective approach that the government repeatedly promised to develop in the wake of the retention and use scandals. The report *Human bodies, human choices* (Department of Health 2002) stated that the following guiding principles would underpin its review and development of legislation:

- respect (for people who have died and their families)
- understanding (that love and feelings of responsibility remain after death)
- informed consent (enabling fully informed choices to be made)
- time and space (to consider what decisions to make)
- skill and sensitivity (in dealing with those close to a patient or deceased person)
- information (to improve understanding and decision-making)
- cultural competence (ensuring that different attitudes to post-mortems, burial, and use of organs and tissue are recognized)
- a gift relationship (shifting the emphasis from organ retention to organ donation).

Developed by the Chief Medical Officer Liam Donaldson, these principles not only allude specifically to respect but, by talking of donation and gift, appear to imply the need for the kind of positive act of will that can only be guaranteed by a strong opt-in system. Similarly, the subsequent Department of Health report *Proposals for new legislation on human organs and tissue* stated that explicit consent would be the fundamental principle underpinning the new legislation (Department of Health 2003:2). However, ultimately it seems that the government was easily persuaded by medical lobbying to support medical use in such a fashion as to hollow out the principle of consent to the point where little more than rhetoric was left. Commin has similarly observed that, for biobanking in France, the notion of informed consent has been emptied of much of its meaning by Article L1211-2, paragraph 2 of the Public Health Code, not least because consent can be disregarded altogether when the donor cannot be found or is deceased (see Chapter 11, section 11.2.1).

Whilst the Human Tissue Act is good for illustrating shortfalls with respect to consent in relation to the deceased (as well as in general),[2] one can also find governance problems elsewhere with respect to other specific classes. For example, general research provisions are not always fully protective of the incapable. This is certainly true of Article 17 (ii) of the CHRB and Principle 24 of the Declaration of Helsinki. However, these provisions should be read down in light of the fact that both these instruments have, as already noted, an overarching orientation towards respect for the individual—one that in the case of the CHRB is specifically re-emphasized in its Protocol on Biomedical Research (not least in Article 15). Nonetheless, they are likely to generate confusion about whether respect for the individual is the critical concern, or at least undermine an appropriate understanding of what it entails. The approach of the Mental Capacity Act 2005 to intrusive research on the incapable adult is a good example of how this kind of confusion can adversely influence national lawmaking processes. One of the pillars of the Act is to protect the interests of the incapable (see particularly section 1(5)) but sections 31(5) and (6) seem to conflictingly point towards the possibility of trading such protection against needs for research (for further analysis see Chapter 2).

[2] However, it is argued by other authors that since human rights protect basic interests of embodied persons, human rights cannot protect cadavers as well as anything that has no interests (see Barilan in this volume), leading to less strict requirements of informed consent and the usage of human tissues and bodily remains.

Use of tissue for research purposes without full protection is already widely evident with respect to *in vitro* embryos. Article 18 of the CHRB prohibits the creation of *in vitro* embryos for research purposes but allows research on existing embryos subject to a proviso of 'adequate protection'. Whether 'adequate protection' is synonymous with 'full protection' must be doubted. Furthermore, whilst some jurisdictions take a more protective approach, a few go as far as allowing embryos to be created for research, including the UK under the Human Fertilization and Embryology Act 1990, which also allows for the creation of cloned and hybrid embryos, in contravention of Article 12 of the CHRB.

3.5 **Respect as a basis for prohibiting commodification**

Commodification of the human body clearly takes place where a provider or user of body material is able to gain money or money's worth from their activity at a level which exceeds due compensation for time, effort, and reasonable costs incurred. Such excess would appear to be precluded by Article 21 of the CHRB which states that the human body and its parts shall not, as such, give rise to financial gain. According to the Explanatory Report, Article 21 'applies the principle of human dignity set forth in the preamble and in Article 1' (para 131). Under it:

> ...organs and tissues proper, including blood, should not be bought or sold or give rise to financial gain for the person from whom they have been removed or for a third party, whether an individual or a corporate entity such as, for example, a hospital. (para 132)
>
> However, technical acts (sampling, testing, pasteurization, fractionation, purification, storage, culture, transport, etc.) which are performed on the basis of these items may legitimately give rise to reasonable remuneration. For instance, this Article does not prohibit the sale of a medical device incorporating human tissue which has been subjected to a manufacturing process as long as the tissue is not sold as such. Further, this Article does not prevent a person from whom an organ or tissue has been taken from receiving compensation which, while not constituting remuneration, compensates that person equitably for expenses incurred or loss of income (for example as a result of hospitalization).

The Article does not apply to 'products as hair and nails, which are discarded tissues, and the sale of which is not an affront to human dignity' (para 133) and most controversially was not considered in relation to, and hence does not yet apply to, issues relating to the patentability and patenting of biotechnological inventions (para 134).

Whilst most jurisdictions have prohibitions on commercial dealing on materials in place, not all accompany these with effective enforcement (in countries like India, for example, a massive black market in kidneys and various other body materials exists). Furthermore, some jurisdictions exclude clearly significant materials such as blood (for example, the USA has a commercial system of blood donation). In addition, extracted material can often give rise to property rights that are not constrained sufficiently to avoid excess (for example, open-ended profit can be obtained from patents for the application of ingenuity to body material). The case of *Moore* v. *Regents of University of California* [1990] 793 Pd 479 is a good illustration of this. John Moore was diagnosed with hair cell leukaemia in 1976. On the clinically appropriate advice of Dr Golde at the UCLA Medical Center of Los Angeles he had his spleen removed. Dr Golde spotted the rare properties and potential commercial value of Moore's blood cells and encouraged him to travel regularly to the Center over the next seven years from his home in Seattle, giving him the impression that these visits and the extractions of various samples that occurred during them were necessarily and solely designed to help ensure his continued health. In fact, Dr Golde and his associate Dr Quan were developing an immortal cell line from his material. By 1990, this cell line was estimated as having a potential commercial value of $3.01 billion. Having found out what had occurred, Moore made numerous claims. By the time the case ended up in the Supreme Court of

California, these had been filtered down to breach of fiduciary duty and a property-based claim, the former of which was accepted as giving rise to a duty to give information about planned further use and the latter of which was rejected.

The eventual outcome of a small settlement on the breach of fiduciary duty point might have been sufficient to pay Moore's legal costs and avoided the dignity-adverse outcome of him profiting from his own material. On the other hand it left the best part of the potential profit intact, thus enabling Dr Golde and the University to gain at a level unrelated to effort. This and other forms of excess have since become thoroughly ingrained in the field of use of tissue for research purposes, not least via the mushrooming of for-profit biobanks and those which operate on a not-for-profit basis but allow commercial exploitation of their data by third parties. Realigning systems away from market power and its Lockean underpinnings will not only be inherently valuable but also remove at least some of the incentives for other types of violation of respect-related norms, such as the breach of trust that occurred in *Moore*.

3.6 Conclusion

Tissue use is becoming ever more systematic, with an increasing amount of it being used for research, not least through the burgeoning development of biobanking initiatives. This chapter has demonstrated the importance of medical use of the body being compatible with respect for the individual and has illustrated some shortfalls of practice and governance related to the use of tissue for research purposes in this regard. It has been observed that 'individual and collective interests are not *per se* mutually exclusive in the field of biobank research' (Chapter 7, section 7.4). Furthermore, on a more philosophical level one might argue that collective interests are only ever ultimately served when they are pursued in a manner compatible with respect for the individual (Garwood-Gowers 2008). However, conceptions of justice, utility, and self-interest that conflict with this philosophy have often been pursued, not least in the context of medical use of the body. Thus whilst those designing, implementing, and regulating programmes of use of tissue for research purposes have a heavy responsibility to guard respect-related norms better, we must recognize that history stands as a warning to the public that they would be wise never to take that guardianship for granted.

References

Bristol Royal Infirmary Inquiry (2001). *Learning from Bristol: the report of the public inquiry into children's heart surgery at the Bristol Royal Infirmary.* CM 5207 (I). Available online at: www.bristol-inquiry.org.uk/final_report/index.htm (accessed 3 January 2010).

Brownsword, R. (2008). Research and new technologies: rights overstated, responsibilities understated. *European Association of Health Law Conference*, 10–11 April.

Calabresi, G. (1991). Do we own our own bodies? *Health Matrix*, **1**, 5–18.

Chief Medical Officer. (2001). *Report of a census of organs and tissues retained by pathology services in England.* London: Department of Health.

Department of Health (2001). *The Royal Liverpool Children's Inquiry Report.* H.C. [Session 2000-2001] 12-I. Available online at: http://www.dh.gov.uk/en/Publicationsandstatistics/Publications/PublicationsPolicyAndGuidance/DH_4009030 (accessed 3 January 2010).

Department of Health. (2002). *Human bodies: human choices.* London: Department of Health.

Department of Health. (2003). *Proposals for new legislation on human organs and tissue.* London: Department of Health.

Dute, J. (2005). The leading principles of the Convention on Human Rights and Biomedicine. In J. Gevers, E. Hondius, and J. Hubben (eds), *Health Law, Human Rights and the Biomedicine Convention*, pp. 3–12, Leiden: Martinus Nijhoff.

Fabre, C. (2006). *Whose body is it anyway? Justice and the integrity of the person.* Oxford: Oxford University Press.

Furness, P. and Sullivan, R. (2004). The Human Tissue Bill. *British Medical Journal*, **328**, 533–4.

Garwood-Gowers, A. (2006). Vindicating the right to bodily security of the incapable in research. *Journal of Mental Health Law*, May, 7–25.

Garwood-Gowers, A. (2008). The right to bodily security vis-à-vis the needs of others. In D. Weisstub and G. Diaz-Pintos (eds). *Autonomy and human rights in healthcare: an international perspective*, pp. 375–397. New York: Springer.

Gunn, M., Wong, J., Clare, I.C.H., and Holland, A. (2000). Medical research on incompetent adults. *Journal of Mental Health Law*, February, 60–72.

Harris, J. (1970). The survival lottery. *Philosophy*, **50**, 81–7

Harris, J. (1997). The ethics of clinical research with cognitively impaired subjects. *Italian Journal of Neurological Sciences Supplement*, **5**, 9–13.

Harris, J. (2002). Law and regulation of retained organs: the ethical issues. *Legal Studies*, **22**, 527–49.

HM Inspector of Anatomy (2003). *The investigation of events that followed the death of Cyril Mark Isaacs.* Available online at: http://www.dh.gov.uk/en/Publicationsandstatistics/Publications/PublicationsPolicyAndGuidance/DH_4064681 (accessed 3 January 2010)

McCullagh, P. (2004). *Conscious in a vegetative state? A critique of the PVS concept.* New York: Springer.

Palazzani, L. (2008). Person and human being in bioethics and biolaw. In D. Weisstub and G. Diaz-Pintos (eds). *Autonomy and human rights in healthcare: an international perspective*, pp. 89–98, New York: Springer.

Price, D.P.T. (2005). The Human Tissue Act 2004. *Modern Law Review*, **68**, 798–821.

Royal College of Pathologists (2001). *Transitional guidelines to facilitate changes in procedures for handling 'surplus' tissue and archival material from human biological samples*, Available online at: http://www.rcpath.org/index.asp?PageID=1086 (accessed 3 January 2010).

Chapter 4

Risky business: re-evaluating participant risk in biobanking[1]

Nils Hoppe

4.1 Introduction

Biobanks are organized collections of biological samples and various forms of associated data, such as medical data, genetic data, and lifestyle information which already are or can be linked to an individual (Cambon-Thomsen 2004:866; Nationaler Ethikrat 2004:9). The question of whether information and data collected and produced on the basis of this material is personal (i.e. can be attributed to a specific individual) carries no weight for the discussions in this chapter for two reasons. First, it is inherent in the research methodology of many biobanks (e.g. UK Biobank 2009) that a linkage to a specific individual is required for the research to be conducted (Campbell 2006:204). Secondly, the nature of genetic data means that the identifiability of the data subject hinges on the density of other data and contemporary technological means (see Arning et al. 2006:701). For the purposes of this chapter, and as a wider position in relation to genetic data, I take the position that these data cannot be anonymized appropriately and permanently.

As instruments of biomedical research, these collections are subject to the same considerations that are appropriate in other kinds of research with human material.[2] The innovative research methods[3] involved and the high volume of information gathered about individuals in the context of biobanks raise new and interesting questions in the fields of ethics and law. The reason for these new challenges lies in the nature of the material and the information stored. The material itself is a data carrier to highly predictive health information, whilst the associated data invariably permit identification of the individual to whom the data belong. This raises a number of issues which deserve more detailed appraisal. When discussing the implementation of innovative methodology in the biomedical setting, it is highly informative to examine how regulation and ethical assessment treat this new methodology.

The inclusion of human participants in research is thought to be ethically inadmissible, illegal, and in need of proscriptive governance if the risk involved for these individuals exceeds the expected benefits. It is easy to envisage situations in biobank research where there is a palpable risk to the participants on the basis of the revealing nature and density of information that these

[1] I am grateful to numerous colleagues and reviewers for their assistance and feedback. As ever, all inaccuracies remain my own.

[2] But not necessarily the same considerations as may be appropriate for research with humans (see Eriksson and Helgesson 2005:1072). I tend to expand Eriksson and Helgesson's distinction: in the instant context, human subjects are, at different times, both directly (at recruitment stage) and indirectly (ongoing research using their tissue and data) involved.

[3] Such as continuous access to health data of participants during the course of a longitudinal study (e.g. UK Biobank 2009:3).

biobanks accumulate. Conversely, the benefits expected by virtue of participation in a biobank project are either on a societal scale (and thus impossible to balance against the risk to the individual) or individual but too remote or ruled out (such as the benefit resulting from incidental findings; see Dressler (2009:85–99) for an exploration of this issue, and see also Chapter 6).

This chapter will discuss two normative functions in relation to the question of risk in biobanking: ethical assessment of research using biobanks, and the regulatory appraisal of research using biobanks. It will be shown that the yardstick by which the notion of risk is assessed differs in these two normative functions, and conclusions will be drawn on the consequences of this discrepancy (e.g. Bentley and Thacker 2004; Directive 2001/20/EC; SCCNFP 2004).

4.2 **Risk as a concept**

The *Oxford English Dictionary* (Simpson and Weiner 1998) defines the meaning of the word 'risk' as '[h]azard, danger; exposure to mischance or peril', firmly putting *risk* in the realm of the undesirable. Shrader-Frechette proposes that the term 'risk' can be used in at least five different ways:[4] (1) the possibility that some harm may occur; (2) the probability of some undesirable outcome; (3) the quantitative probability that some consequence will occur; (4) the cost of some negative occurrence; (5) the chance of loss occurring (Shrader-Frechette 1998:331). These categories do not appear to be exclusive, and a certain degree of overlap is evident. Categories (1) and (5) contain an element of possibility and chance. This simply conveys the uncertainty of some sort of mischance or peril. Categories (2) and (3) suggest a quantitative notion of risk. Category (4) addresses risk as a magnitude of loss. Therefore risk can be cast in terms of either a *probability* or an *uncertainty* of a detrimental event occuring. *Probability* signals that the concept is accessible by means of quantitative analysis ('the risk of rain is 15 per cent'). *Uncertainty* characterizes the concept as incalculable and undetermined ('there is a risk that it will rain today').

4.2.1 **Types of risk**

In the context of biobanks, the types of risk that are discussed are quite diverse. It seems clear that they go over and beyond the individual risk of actual physical harm which may materialize at the recruitment stage.[5] Additional harm may result from information which has been gathered from the research participant falling into the wrong hands (Eriksson and Helgesson 2005:1071) as well as subsequent discrimination or stigmatization. The combination of the data and the material provided, as well as the results of subsequent analysis, along with the exchange of the information in networks and databases of varying security adds another dimension to the context. Categories of harm are suggested to include *physical*, *psychological*, and *social* harm (van Luijn et al. 2006), and possibly also *moral* harm (Eriksson and Helgesson 2005). It is opportune at this point to pose the question as to whether these categorizations are sufficiently abstract to be useful in times of rapid biotechnological advances. New research methodologies based on the revealing character of genetic data emerge frequently. The indefinite, ongoing, and unilateral procurement of lifestyle

---

[4] Shrader-Frechette suggests that moral philosophers use the term in these five different ways. It is clear that these categories are not restricted only to moral philosophers.

[5] Such as where a participant has an adverse reaction to the method of excising material or an adverse needlestick reaction. A recent study (on venipunctures in the context of blood donation, which entails giving a substantially larger sample) suggests that, in *probability* terms, the risk of an adverse reaction is reasonably low (1.2 per cent; $n = 4906$) (Crocco and D'Elia 2007). The variety of different reactions classed as adverse in this study range from very mild (agitation, sweating, pallor, cold feeling, sense of weakness, nausea) to severe (vomiting, loss of consciousness, and convulsive syncope).

and health information to accumulate an unheard of density of data about an individual's health carries consequences that have as yet not been discussed to a sufficient depth to yield the type of knowledge we might need to assess the related risks. The very nature of this research, as pointed out, also precludes any reasonable semblance of anonymization and thus permanent protection of individualized information.

4.2.2 **Individual risk or societal risk**

The prime exercise in relation to risk is balancing the ratio of risk versus benefit. The question whether the risk is one which applies to an individual or a group of individuals influences the nature of this balancing exercise. Depending on the context, this distinction may prove crucial.[6] Particularly where the gamut slides from individualized consequences to consequences of significance for public health, the approach to assessing risk changes dramatically. Any formula we establish falters when we no longer balance benefits and risks only in terms of individuals. In some instances, risk to an individual is balanced against a benefit to society, rather than a benefit to the individual at risk.[7] Conversely, it is quite difficult to envisage a situation where a risk to society is acquiesced to for the benefit of an individual. Simply put, it is necessary to bear in mind that eligible benefits may include those which do not accrue to the individual whose risk we are assessing. The public interest in undertaking certain research or a risk in the context of public health can act as a manipulator, tipping scales in favour of running a certain risk.

4.2.3 **Who defines risk?**

As set out above, the term 'risk' can be characterized in different ways.[8] The question of individual and societal perspective will play an important role in this characterization. This is reflected in the question of who defines the acceptable level of risk. This takes on specific significance when put into the context of the subject matter of this chapter: whether different normative functions carried out on the basis of the idea of risk use the same principles and presumptions. For example, taking the concept of 'minimal risk', as enshrined in US federal legislation,[9] raises the question of its utility for other protagonists. What do ethics committees and courts make of the policy-makers' idea of 'minimal risk'? Further, some types of risk are inherently subjective: the risk involved in being given a blood transfusion when ending up in hospital carries different subjective weight for a Catholic than it does for an orthodox Jehovah's Witness. Additionally, there is some evidence that individuals are happy to run highly probable risks which carry fes consequences, whilst they are reluctant to run risks of very low probability which are associated with catastrophic consequences (Luhmann 1991:3). In the context of biobanks, the decisive question is whether traditional concepts of risk still hold water when there is more at stake than just the risk associated with a biopsy. What about the subsequent risk of genetic stigmatization or discrimination, of exposure to forensic examination, and loss of liberty?

---

[6] The same is true for the evaluation of benefits that may accrue on either an individual or a societal level (for a more detailed elaboration of this issue, see Chapters 5, 6, and 7).

[7] It might be said that the responses accumulated by van Luijn et al. point to such thinking. Whilst the benefit for the patients was thought to be minimal or trifling, the research was felt to be desirable. Clearly, in this case, the desirability was a parameter of benefit to society rather than to the patients (van Luijn et al. 2006:172).

[8] An example of different risk perceptions in the context of autologous stem cell applications in the countries of Germany and the UK is given in Chapter 16.

[9] 45 CFR 46, §404.

Additionally, it becomes clear when viewing the normative functions based in assessments of risks (such as the work of institutional review boards (IRBs) or courts) that the point in time when such assessments take place significantly influences the assessment itself: IRBs will attempt to assess the risk ex ante, whilst courts tend to deal with risk ex post—when it has already materialized.

4.2.4 Risk as a 'probability' or an 'uncertainty'

At first glance, the question as to whether the terms 'probability' and 'uncertainty' can be used in different ways may seem trivial. A closer look reveals that the terms may lend themselves to distinguishing two approaches to determining risk as described above. The uncertainty of risk appears to match an approach based on informal assessment of risk, whereas the notion of 'probability' seems to tally with a more formal quantitative assessment of risk. The two terms are not used in exactly the same way in ordinary language, and this distinction may be enlightening. These reflections distil two different issues: first, whether an assessment is based on informal qualitative parameters or on formal quantitative parameters, and, secondly, at which point in time risk is actually assessed—when it is an uncertainty or a probability, or when it has already materialized and is a certainty.

4.3 Contributory parameters

Having briefly investigated the nature of the concept of 'risk', it is time to examine the concrete subject matter. As indicated at the outset, the nature of harm operates on a sliding scale from actual harm involved in a venipuncture at recruitment stage all the way to subsequent information-related harm. These different categories of harm are usually separated into *physical* and *non-physical* harm (Eriksson and Helgesson 2005:1072), or *physical*, *psychological*, and *social* harm (van Luijn et al. 2006:172). Eriksson and Helgesson also introduce the concept of *moral harm* where there is no damage in the legal sense. This covers going against the participant's express wishes in using or storing the material. In the following, I will briefly raise issues in relation to how the origin and the type of material or data influence the notion of risk before turning to questions of consent and commerciality.

The rationale behind investigating the origin of material lies in whether additional harm is incurred by taking the material which is to form part of the collection. Where material is harvested purely for research purposes, any risk that may be acquiesced to is put in relation to the benefit to be expected from participation in that research. In the context of large-scale population-based biobanks aimed at recruiting healthy volunteers, the balancing exercise is simply impossible. If a counterweight is needed, it may be found in a palpable benefit to society rather than to the individual. Where the material procured is made up of surgical or diagnostic leftover material, the balance is between the risk of the intervention and the benefit expected from the surgery or the diagnostic method, not the research.[10] This assessment applies only to the risk involved in obtaining material; in cases where the material is obtained for diagnostic or other primary purposes and only later selected for inclusion in research, it is simply not possible to consider questions of the risk of participating in research at the point of procurement. An additional dimension arises in situations where there may be an individual risk assessment for each of the following: procurement of material, information and genetic analysis, or other tests. This kind

[10] See also Chapter14 for a discussion of using pathology archives for research.

of risk assessment cannot extend to the risk potential developing upon combination of all of these constituent parts.

It is evident that the kind of material taken plays an important role in terms of both the invasiveness of the excision and the moral charge of the material itself. It is well outside the remit of this chapter to go into detailed discussions of this issue. A detailed assessment of the significance of varying the type of tissue and the kind of research done on the admissibility of this research can be found elsewhere (Hoppe 2009:14–25). For the purposes of this chapter it is sufficient to restate that the kind of material with which the research is conducted is of particular relevance. From differences between hair and nail clippings to a biopsy sample and gametes we can extend the analysis to the kind of data that is attached to the material. Some participants may not mind giving data in relation to their alcohol and tobacco consumption which will be included in the biobank's data dossier on them. However, they may take issue with the inclusion of information on sexually transmitted diseases or psychiatric illnesses. Therefore the type of material and data determines the exact nature of risk, in both objective and subjective terms.

Finally, issues of commerciality are of immense significance. The provision of payment or financial reward may influence the willingness of potential participants to take on risks. Bentley and Thacker (2004:297) suggest that their study, which investigated the influence of pecuniary incentives on risk-taking, reveals '...that monetary payment increases respondents' willingness to participate in research regardless of the level of risk'. Therefore the suggestion is that the payment of research participants influences the subjective assessment of risk. This leads to a softening of the impact of criteria such as 'probability' and 'uncertainty'. When seeking to establish whether consent obtained from research participants has been unduly tainted by the promise of pecuniary profit, it is necessary to discuss openly whether cash for participation is an undue influence or simply another kind of benefit that needs to be taken into account. It is the issue of consent which carries the most dogmatic weight when discussing risks from a legal perspective.[11] It is widely accepted that it is simply not possible to consent to all kinds of activity, particularly those which carry a high probability of severe detriment.[12] This gives an indication as to the weighting of some potentially exculpating parameters such as consent. The provision of reimbursements or payment for running a risk does not change that particular threshold, which raises the question of whether a simple economization of risk-taking is per se undesirable.

4.4 Approaches in ethics and in law

We have seen that different normative functions exercised by different groups or entities (such as IRBs or courts) have different approaches to viewing risk whilst at the same time applying the notion of 'risk' in a normative fashion when assessing the admissibility of research. In the following, I will cite examples of such use and will draw comparisons and outline discrepancies to facilitate a first appraisal on the usefulness of these risk concepts.

4.4.1 Ethics

Biomedical research is exceptionally well placed to cause both a great deal of benefit and intense detriment to individuals, communities, and society, simply on the basis of its everyday work. Ethics is concerned with moral reflection on these kinds of activities and with the identification of behavioural paths which can be deemed morally admissible in differing circumstances.

[11] In addition, the extent and quality of the kind of consent needed in this context is by no means beyond contention (see Chapter 2 for a discussion).

[12] See *R* v. *Brown* et al. [1993] 2 All ER 375 for an unconvincing but authoritative ratio to this effect.

These determinations often form the basis of subsequently enshrining proscription or prescription in law or of informing policy at a societal level. In this capacity, the discipline of biomedical ethics takes a position of eminent importance, particularly in the field of research with human subjects.

Whilst such research is fraught with difficulty, it is also an extremely desirable activity. Without excellent medical research, the availability of the best possible care for future patients would be undermined (for a broader argument and the appropriate sources to this effect see Hoppe 2009:29). Biobanking includes a varied range of potential risks, from adverse reactions at recruitment to genetic discrimination. Nonetheless it is quite clear that, as an instrument of innovative epidemiological research, biobanks are indispensable. Having established that a certain perilous activity is, by and large, wanted, the next step is to investigate how the activity might be carried out in the most appropriate way. It appears clear that an unsafe activity does not merit the label of moral inadmissibility per se. Radcliffe Richards argues convincingly:

> ...when you find real dangers and difficulties, you will try to devise techniques for avoiding the harm while keeping the good. This is, after all, what we automatically do in other contexts. The existence of rogue builders and medical quacks does not lead us to try to stop building or medicine altogether: obviously, we aim for controls that will minimise the bad and keep the good. (Radcliffe Richards 2003:140)

Such techniques and controls are appropriate protocols for assessing the risk–benefit relationship of certain research. The usual course of action in order to establish whether something is morally objectionable in this context is a balancing exercise between the benefits and the associated costs or risks (Beauchamp and Childress 2001:194). Beauchamp and Childress compellingly suggest that in practice the techniques used by IRBs to ascertain levels of risk are typically restricted to informal techniques (such as expert opinions and analogy by precedent). In contrast, they go on to suggest formal quantitative methods of ascertaining risks (ibid.:194–206). The application of the latter is an expression of probability, whereas an IRB's application of informal techniques points to uncertainty. In both types of assessment, one is concerned with producing some sort of finding in relation to the cost of a risk materializing. It is also determined how likely or probable the materialization of the risk is. The risk run has to be set side by side with any benefit involved in undertaking the activity as a simple trade-off between the good and the bad. Only where the benefit outweighs the risk may we seriously consider the activity.

> Every medical research study involving human subjects must be preceded by careful assessment of predictable risks and burdens to the individuals and communities involved in the research in comparison with foreseeable benefits to them and to other individuals or communities affected by the condition under investigation. (WMA 2008:para. B.18)

Similar provisions can also be found in the Oviedo Convention:[13]

> Research on a person may only be undertaken if...the risks which may be incurred by that person are not disproportionate to the potential benefits of the research [.] (Chapter V, Article 16(ii))

These lines have legislative quality for the jurisdictions which have ratified the Convention. Similarly, the generally non-binding character of the Declaration of Helsinki is given indirectly

[13] Convention for the Protection of Human Rights and Dignity of the Human Being with regard to the Application of Biology and Medicine: Convention on Human Rights and Biomedicine of the Council of Europe of 4 April 1997. The Oviedo Convention is not in force in all Member States of the European Union. Notable absentees from ratification are Belgium, Finland, France, Germany, Republic of Ireland, Italy, Latvia, Luxembourg, Malta, Netherlands, Poland, Sweden, and the UK.

binding legislative quality where licensed members of the health professions are required to adhere to its stipulations or risk being struck off. There is a convincing argument that the risk–benefit balancing exercises outlined in these two examples are directed at ethics committees and have normative quality. Thus, at first sight, the task for IRBs seems to have been reduced to a cost–benefit analysis. Some commentators suggest that risk–benefit analyses constitute the principal activity of IRBs (van Luijn et al. 2006). The challenge lies in the method used to reach an understanding of how certain or how probable it is that the risk materializes. Informal methods extend to understanding the risk on an analogy base (i.e. how has the risk materialized in the past; what do experts think about the possibility of the risk materializing), whereas formal methods seek to establish a quantifiable probability. Those tasked with measuring the risk versus benefits may find it difficult to pinpoint a methodology to do so (van Luijn et al. 2006:170). Nonetheless, van Luijn et al.'s study of how members of IRBs weigh risks and benefits to reach certain decisions reveals some significant features of ethical risk reviews. The risks discussed in this study included physical, psychological, and social burdens. The range of possible benefits included a significant range of clinical indicators (such as tumour remission, less toxicity, etc.), but the greatest weighting (65 per cent) was given to the mitigating benefit indicator 'hope' (ibid.:174). Unlike tumour remission and toxicity, the impact of hope is clinically impossible to measure. This means a shift from quantitative formal indicators to informal indicators. The authors also conclude that 'the results...indicate that, while members of IRBs believed the research to be important, they expected only modest benefits to accrue to the participant patients.' (ibid.:172). There might only have been a modest benefit to the individual, but the importance of the research to the scientific community or to society is used to manipulate the balancing exercise. It is brought into play to tip the scales in favour of research, despite the modesty of expected benefits.

A final aspect which immediately becomes apparent when looking at how risk assessments are undertaken in this context is their predictive quality. Risk assessments in the context of entities such as IRBs aim at identifying ex ante whether a research undertaking that requires participants to run a certain risk is to go ahead or not. This explains the mixture of formal and informal mechanisms, attempting to balance hard evidence of probability with notions of uncertainty. This becomes particularly significant when compared with the legal context in which risk is assessed.

4.4.2 **Law**

Whilst IRBs and ethics committees will include legal reasoning in their decision-making processes (for a comparison of ethics committees in 10 European countries, see Hernandez et al. 2009) there has been some debate on the judicial nature of such bodies (e.g. Hendrick 2001). Some aspects of the work of ethics committees have been enshrined in items of regulation, and it is here that we can look for an illustration of how legal reasoning might interface with ethical decision-making (for an elaboration of the relationship between ethics and law, see Chapter 8).[14] In the UK, the Medicines for Human Use Regulations[15] require ethics committees to ascertain whether an appropriate risk–benefit assessment has been carried out. Schedule 1, part 2, section 2 states:

> Before the trial is initiated, foreseeable risks and inconveniences have been weighed against the anticipated benefit for the individual trial subject and other present and future patients. A trial should be initiated and continued only if the anticipated benefits justify the risks.

[14] For example in section 15 of the Medicines for Human Use (Clinical Trials) Regulations 2004 (Statutory Instrument 2004 No. 1031) (the Medicines for Human Use Regulations).

[15] S.I. 2004 No. 1031.

Here we have a regulation containing guidance for ethics committees, encapsulated in legal language. The standard of relevant risk is reduced to one of the *foreseeable* risk. The array of different contexts in which the law has had to address the question of risk ranges from foreseeing harm materializing in terms of medical negligence[16] to the precise (or indeed not so precise) type of risk which ought to be disclosed to patients for the purposes of obtaining sufficiently robust consent.[17]

> [A doctor] must take care in the examination, diagnosis and treatment of his patient's condition to prevent injury to his health from risks which a competent practitioner would foresee as likely to result from his failure to do so. (*Brown* v. *Lewisham and North Southwark Health Authority*, per Beldam LJ)

The approach of courts to the notion of 'risk' appears to be altogether more pragmatic and with a sharper outline than that deployed by IRBs. A plausible connection of concepts made here is not whether a risk is simply too high compared with the benefits but whether a *reasonable person* would have accepted the risk under the circumstances (Stauch 2001:195). This concept does not provide an answer in cases where the risk has materialized but would have been run anyway by a reasonable person as the standard is shifted to a subjective perspective. What it does provide is an answer to the question of liability, which shifts the temporal context of the risk assessment. A statement in relation to the foreseeability of a certain risk materializing is made after the risk has materialized and blame or responsibility is to be apportioned to someone. This constitutes the second part of the comparison with ethical decision-making. In ethical decision-making, one is concerned with predictive analysis of risk; in law, risk is dealt with ex post. From the rules developed after the event to ascertain whether a risk *should have been run* (not *should be run*), legislative guidance including elements of foreseeability is developed, purportedly for the benefit and guidance of ethics committees who work ex ante. Whether the circularity of this exercise is appropriate is very doubtful.

4.5 **Conclusion**

The brief discussion of different characteristics of the notion of risk in this chapter has shown that the concept is by no means as easily captured and as normatively coherent as it may appear. A number of extrinsic factors influence the quality of the idea and render it cumbersome and in some cases inappropriate. It is a step too far to consider *research using human tissue and data* in the same way that we might consider *research using humans* (Eriksson and Helgesson 2005:1072). The potential harm attaching to direct participation in research (taking experimental substances or undergoing regular intervention and tests) is clearly different to the type of harm that we might expect to arise in relation to providing material and information for research. Categories of harm may need to be reconsidered in the light of new future risks arising when several sources of data on individuals are combined. Most importantly, it is clear that the utility of using the concept of 'risk' in the ethical setting and the legal setting without questioning its coherence is limited. Institutional review boards are concerned with an *ex ante* determination of risk whereas courts

---

[16] For example, *Brown* v. *Lewisham & North Southwark Health Authority* [1999] EWCA Civ 1063 where, six days after a quadruple coronary artery by-pass, the plaintiff was to take public transport back from the hospital carrying out the operation to his admitting hospital in Blackpool. During the course of the journey, he developed an unrelated thrombosis which led to the loss of a leg. He claimed that the thrombosis would have been spotted had he still been in hospital.

[17] *Sidaway* v. *Board of Governors of the Bethlem Royal Hospital* [1985] UKHL 1; *Chester* v. *Afshar* [2004] UKHL 41.

have the benefit of hindsight when working with 'risk' as a possibly foreseeable certainty. Whilst no immediate recommendation is forthcoming as part of this chapter, it is clear that, when working with the concept of risk, it is prudent to bear in mind the interdependencies of by no means coherent parameters and settings—particularly whether principles and guidance were designed for ex ante or ex post evaluation of risks.

References

Arning, M., Forgó, N., and Krügel, T. (2006). Datenschutzrechtliche Aspekte der Forschung mit genetischen Daten [Data privacy: legal aspects of research with genetic data]. *Datenschutz und Datensicherheit*, **30**, 700–5.

Beauchamp, T.L. and Childress, J.F. (2001). *Principles of biomedical ethics*. New York: Oxford University Press.

Beck, U. (1986). *Risikogesellschaft—Auf dem Weg in eine andere Moderne [A society of risk—on ist way to a new modern age]*. Frankfurt: Suhrkamp.

Bentley, J.P. and Thacker, P.G. (2004). The influence of risk and monetary payment on the research participation decision making process. *Journal of Medical Ethics*, **30**, 293–8.

Bonß, W. (1995). *Vom Risiko: Unsicherheit und Ungewissheit in der Moderne [Of risk: insecurity and uncertainty in the modern age]*. Hamburg: Hamburger Edition.

Cambon-Thomsen, A. (2004). The social and ethical issues of post-genomic human biobanks. *Nature Reviews Genetics*, **5**, 866–73.

Campbell, A.V. (2006). The ethical challenges of biobanks: safeguarding altruism and trust. In S.A.M. MacLean (ed.), *First do no harm: law, ethics and healthcare*. Aldershot: Ashgate.

Crocco, A. and D'Elia, D. (2007). Adverse reactions during voluntary donation of blood and/or blood components. A statistical epidemiological study. *Blood Transfusion*, **5**, 143–52.

Dressler, L.G. (2009). Biobanking and disclosure or research results: addressing the tension between professional boundaries and moral intuition. In J.H. Solbakk, S. Holm, and B. Hofmann (eds), *The ethics of research biobanking*. Dordrecht: Springer.

Eriksson, S. and Helgesson, G. (2005). Potential harms, anonymization, and the right to withdraw consent to biobank research. *European Journal of Human Genetics*, **13**, 1071–6.

Hendrick, J. (2001). Legal Aspects of Clinical Ethics Committees. *Journal of Medical Ethics*, **27**, i50–i53.

Hernandez, R. et al. (2009). Harmonisation of ethics committees' practice in 10 European countries. *Journal of Medical Ethics*, **35**, 696–700.

Hoppe, N. (2009). *Bioequity—property and the human body*. Farnham: Ashgate.

Luhmann, N. (1991). *Soziologie des Risikos [A sociology of risk]*. Berlin: de Gruyter.

Nationaler Ethikrat. (2004). *Biobanken für die Forschung [Biobanks for research]*. Berlin: Nationaler Ethikrat.

Radcliffe Richards, J. (2003). Commentary: an ethical market in human organs. *Journal of Medical Ethics*, **29**, 139–40.

SCCNFP (Scientific Committee on Cosmetic Products and Non-Food Products Intended for Consumers) (2004). Recommended mutagenicity/genotoxicity tests for the safety testing of cosmetic ingredients to be included in the annexes to Council Directive 76/768/EEC. (SCCNFP/0755/03). Available at: http://ec.europa.eu/health/ph_risk/committees/sccp/documents/out273_en.pdf (accessed 8 December 2009).

Shrader-Frechette, K. (1998). Risk. In E. Craig (ed.), *Routledge Encyclopedia of Philosophy*, Vol. 8, pp. 331–4. London: Routledge.

Simpson, J. and Weiner, E. (ed.) (1998). *Oxford English Dictionary* (2nd edn). Oxford: Oxford University Press.

Stauch, M.S. (2001). Risk and remoteness of damage in negligence. *Modern Law Review*, **64**, 191–214.

UK Biobank (2009). UK Biobank information leaflet. Available at: http://www.ukbiobank.ac.uk/docs/CAT2INFOV001.pdf (accessed 9 December 2009).

van Luijn, H.E.M., Aaronson, N.K., Keus, R.B., and Musschenga, A.W. (2006). The evaluation of risks and benefits of phase II cancer clinical trials by institutional review board (IRB) members: a case study. *Journal of Medical Ethics*, **32**, 170–6.

WMA (World Medical Association) (2008). *Declaration of Helsinki*. Available at: http://www.wma.net/en/30publications/10policies/b3/index.html (accessed 8 December 2009).

Chapter 5

Reciprocity, trust, and public interest in research biobanking: in search of a balance

Nadja K. Kanellopoulou

5.1 Introduction

Research interest in understanding human genetic disease inheritance and its clinical and environmental factors dates back a few decades. In recent years, large-scale biobanks have been set up to research disease pathways in common complex diseases. They investigate patterns of disease inheritance and the interaction of genes, lifestyle, and environment. Genomic research initiatives generate rich phenotypic and environmental datasets in the study of genotypic information, as one of many significant variables to help explain health outcomes. Large numbers of volunteers are needed for these resources, and interest is high amongst researchers to maintain long-term relationships with participants. These changes are matched by a new emphasis on ethical, legal, social, and economic issues in managing research on this scale. They raise fundamental questions about balancing the interests of research participants, the research community, and society, maintaining public legitimacy for such projects, and building sustainable governance frameworks.

The increasing range of national and regional projects can be divided into broad categories: (a) initiatives funded by national governments, based in academic and medical departments, in cooperation with commercial enterprise (Estonia, Scotland, Sweden); (b) government-funded institutes set up for research purposes with no commercial involvement (Norway, Quebec, Singapore); (c) commercial companies backed by the pharmaceutical industry and licensed by the government for drug discovery and development (Iceland—now bankrupt; Tonga—now defunct). Each project has a distinct remit, structure, and funding strategy, and their scope, research design, and recruitment options vary considerably within distinct social, economic, and cultural contexts (Cambon-Thomsen 2004:868).

These projects collect, store, and analyse DNA samples and data according to protocols that allow linkage with medical records and personal information collected in interviews, as well as with genealogies and family histories. The possibility of linkage, access to health records, and analysis based on medical family history research raises complex questions for participants' trust. A plethora of controversial issues beleaguers the development of such large-scale projects. In this chapter, I focus on issues of control, trust, and managing expectations, and draw on two UK-based projects. The UK Biobank and Generation Scotland vary in scale, methodology, and research approach, but were both established to serve the public good (see Chapter 6, section 6.2).

5.2 Major UK initiatives

5.2.1 The UK Biobank

The UK Biobank is funded jointly by the Medical Research Council (MRC), the Wellcome Trust, and the UK Department of Health as a resource for research on the effects of genetic and environmental risk factors of common multifactorial diseases affecting adults in the UK. The project aims to collect biological samples from 500,000 healthy volunteers, aged between 40 and 69, including height, weight, and blood pressure measurements, details of medical history, and lifestyle information. The team will monitor follow-ups on medical and other health-related records in succeeding years. Researchers will apply for access to the UK Biobank to study the complex environmental and genetic factors of common serious disorders such as heart disease, stroke, cancer, and diabetes (UK Biobank 2009).

A series of surveys, polls, and working papers were commissioned between 2000 and 2003 to encourage public support for the project, with emphasis on future public health benefits, ethics, and competent management, and extensive use of a language of genetic citizenship, solidarity, and altruism. The promises of future health and wealth were set against a background of public trust crises in both the medical and scientific spheres, widely reported by the media. At that time, law and ethics scholars highlighted the 'crisis in confidence' and lack of public trust in science in the UK, and the need for genuine consistent dialogue (Laurie 2002:309; O'Neill 2002:11). Social scientists cautioned against importing of notions of altruism and donation from other areas of research when regulating public participation in human genetic databases (Busby 2004:39; Tutton 2004:19).

5.2.2 Generation Scotland

The quest for better understanding of key diseases burdening the Scottish population (such as cancer, heart disease, stroke, mental health) is driving an umbrella initiative between the Scottish Universities of Aberdeen, Dundee, Edinburgh, Glasgow, and St Andrews, the MRC Human Genetics Unit, the National eScience Centre, the Scottish School of Primary Care, and National Health Service (NHS) Scotland. Generation Scotland (GS) is a collaborative initiative funded by the Scottish Funding Council to promote research that can help evaluate healthcare and economic implications of genetics (Generation Scotland 2009). The project is recruiting individuals from different Scottish regions to build representative control cohorts of Scotland's populations and create a knowledge base of genetic information for current and future studies.

The first major study undertaken by GS, the Scottish Family Health Study (GS:SFHS), aims to recruit 50,000 family members aged between 35 and 55 in a large family-based cohort. Two unique characteristics of GS:SFHS are a) the large general population family-based nature, and b) the ability to link individual data with detailed past and future health records. In this project, family recruitment becomes a necessary component of success; effective family management is paramount, with families considered as valuable resources in the study of genetic disease inheritance, especially in common complex diseases affecting regional populations where health problems cluster in families. The research team considers the project important also because of inclusion of detailed cognitive function phenotype information, and collection of data on serious mental illness that could provide opportunities to study the genetics of these areas (Smith et al., 2006:80).

Understanding the reasons why family members may be willing to share personal information is important since research of this kind can have implications for family relationships. Family-targeted databases are not entirely new in the UK; the Avon Longitudinal Study of Parents and

Children (ALSPAC) and Oxagen, which are small disease-based biobanks, have been in operation since the 1990s but with a different structure and purpose. Issues arise in representation and consent for families because novel ways are devised to recruit family members. For example, arrangements are made for a family member, or a person separate from the family, to contact others so that family members do not know who has a particular condition or who has provided their name as a contact. Additional concerns can be the potential for social stigmatization if the impression is created that a person who is taking part in the study (or his/her relatives) is affected by a genetic condition. Challenges emerge in understanding differences in the motivation and expectations of family members and the general public about whether or not disease exists within a family; family members may be more likely to participate in the research if they feel that they will have more to gain from research outcomes, seen as short-term benefits from future research. This particular issue raises ethical questions about the vulnerability of family members that can make them susceptible to overvaluing anticipated benefits. Such considerations highlight the significance of perceptions that may affect people's willingness to take part in research (see Chapter 6, section 6.2).

Assessment of the array of concerns raised by biobanks is beyond the scope of this chapter, especially since these range across a wide spectrum of issues including recruitment and consent, feedback and withdrawal, security, privacy and access, control, commercialization, benefit sharing, ownership, and ethical oversight. At the core of these debates lie fundamental assumptions about public participation in research, linked to broad questions as to whether biobanking projects represent a valid use of public resources and are worthy of public support, about unforeseen exclusionary or discriminatory effects that may incur, about who is likely to benefit from future research, and issues of public or private ownership in such resources (Petersen 2007:32).

5.3 **Ethical principles**

Some countries have responded to public and scholarly concerns by establishing specific legislation or independent ethics governance bodies for increased scrutiny over biobanking practices and standards. The Human Genetics Commission in the UK recommended the principle of *respect for persons* as a general ethical basis for developing protection in the regulation of genetic information. This principle requires biobanks to 'acknowledge the dignity of others and…treat them as ends in themselves and not merely instrumentally as means to ends or objectives chosen by others'. The Commission identified consent, privacy, confidentiality, and non-discrimination as secondary principles to this goal (HGC 2000:52; HGC 2002:35). Traditionally, these principles are implemented through consent mechanisms. The UK Biobank policy on access principles recognizes that 'access [to the resource] is to be managed as to protect participants, honour commitments made to them, act within the scope of their consents' (UK Biobank 2009). Sole reliance on consent has its problems. It is not sufficient to guarantee ongoing control for participants in the use of large-scale resources over time, nor does it provide any means to effectively designate what happens to samples after use or to how potential benefits can be distributed. At the same time, advisory guidance in the UK suggests that participation in genetic research should be deemed a free gift, an assertion which raises critical questions about long-term sustainability of existing biobanking frameworks.

Current criticisms of the adequacy of consent to address the interests of potential participants in meaningful control over the use of samples and data that they provide for research focus on a logic of *disempowerment*, imposed by the traditional view of consent as a merely informational process. Critics view consent as a device that allows the entrenchment of power distribution in the relationship between researchers and participants in ways that fail to make the latter less visible and frame their *relationship* in unequal terms. In this view, participants are being denied a role in controlling the use of possible research resources and benefit, while consent becomes a slim legitimization device in the absence of broader visions on research governance (Laurie 2002:309).

In an era of increased economic and commercial views of human tissue, concerns about entrenched inequality in the ways in which the relationship between participants and researchers is traditionally viewed obtain new significance. This concern is manifest in scholarly attention to questions of whether participants should be provided with information about commercialization when considering involvement in research, or whether they should be allowed to seek a direct or indirect benefit from it, either for themselves or on behalf of their family or the broader community. This issue is increasingly debated among stakeholder groups and in public consultations in the UK, where it has been revealed that prospective participants want to be informed about the financial and commercial aspects of biomedical research (Shickle et al. 2003:1).

This issue merits attention because of policy concerns on how the prospect of commercialization affects the decision to participate in research. It fits with general public concerns about commercialization and the ways in which it may affect investment in good research, but it is also due to ongoing efforts by patient and advocacy groups to protect their collective interests in influencing research direction and use (Terry et al. 2007:160). The emergence of patient interest groups and other collectives in this area encourages a *social contract* view of researcher–participant relationships (Kanellopoulou 2009a:209). Several scholars argue that the way forward is to empower participants to take a *more equal role* in the research partnership (Laurie 2002:309; Winickoff 2007:440). Some insist that, in the absence of provisions for compensation or commitment that take account of participants' concerns, individuals or groups who become aware of the potential for commercialization of research results or outputs may be discouraged from giving biological material to research, with destructive implications for the future sustainability of long-term endeavours and biomedical research at large.

These tensions emerge in a climate of increased realization that the value of human tissue has changed in the last decades (see Chapter 9, section 9.2). At a scholarly level, this awareness is mirrored by a renewed understanding of the status transformation of human tissue from gift to commodity as a form of circulating value (Appadurai 1986:11), which compels a re-evaluation of the seemingly mutually exclusive relationship between gifts and commodities (Waldby and Mitchell 2006:24). However, current legal frameworks do not accommodate the granularity that social sciences analysis brings to the discussion. This absence creates difficulties and engenders ambiguities and incongruities such as importing the blood donation model into research biobanking. It may no longer be sustainable to regulate this field on the basis that people do not see value in 'their' bodies and do not retain an interest in how samples are used or what happens to them after use.

5.4 **Public expectations and anxieties**

A number of public anxieties in this context can be categorized as questions about (a) the power of participants to influence, control, and represent their interests in determining how samples and future research should be used, (b) how research-related benefits can be assessed, (c) how returns can be realized, and (d) the extent to which potential profits could be fed back to the resource or to one's community, in particular in cases where commercial access is being considered. These raise two strands of concerns related to research benefits and overall willingness to participate in research.

5.4.1 Benefits

In 2000, the Human Genome Organisation (HUGO) Ethics Committee suggested that '...a benefit is a good that contributes to the well being of an individual and/or a given community...' (HUGO Ethics Committee 2000). In accordance with the broad consensus that the Committee statement indicates, benefits transcend avoidance of harm (non-maleficence) in so far as they promote the welfare of an individual and/or a community. Benefits are not identical with profit in the monetary or economic sense; determining them depends on needs, values, priorities, and cultural expectations. In research biobanking, the calculation of research benefits and associated returns is rather vague and complex. Profits and benefits can be uncertain, contingent, realized several years after the collection of biological samples and relevant information, or even not realized at all (Rothstein and Griffin Epps 2001:228; Nightingale and Martin 2004:564). This residual uncertainty can cause difficulties in achieving an appropriate balance between public interest in research investment, participants' expectations, and mechanisms for distribution of benefits that may or may not be derived from research. It further nurtures a series of important questions that have yet to be considered systematically in existing policy, perhaps partially because empirical research in this area is still developing. These include whether health and wealth benefits are mutually exclusive, how people perceive anticipated benefits, especially in different cultural and ethical contexts, to what extent these concepts link with how people view their contribution to research, and therefore how the potential for commercialization of research ultimately shapes people's willingness to participate in research.

5.4.2 Willingness to take part in research

According to surveys in the aftermath of the Alder Hey scandal in the UK, people showed a general appreciation of the benefits of science and technology research in society, together with a tension between perceived benefits and a sense of society not fully in control of the pace or direction of research, and a perceived lack of openness and accountability (Cragg Ross Dawson 2000:13). These notions also draw on perceptions of scientists who can have too narrow focus on their work and fail to take account of wider ethical guidelines and priorities within society as a whole (Office of Science and Technology and Wellcome Trust 2000:24). Such views also relate to a lack of knowledge about how science and medical research operate together. Views on genetic research in particular vary from a substantial hope for the potential to provide treatments and therapies to significant concerns about unethical research and some agreement that genetic information requires particular protection (HGC 2000:78). Some people tend to be more accepting of genetic research when they are familiar with genetics because of previous illness, or because of a lack of fear (Cragg Ross Dawson 2000:21).

Several studies have looked at anticipated willingness rates in biobank research in UK Biobank consultations. These vary according to types of tissue, research purpose, and nature of safeguards in place. These consultations also looked at factors that motivate people to take part in biobank research. A general typology of willingness to participate was divided into people who (1) are keen to participate, (2) analyse the costs and benefits of participating (to themselves and others), (3) are passive or see no reason not to participate, and (4) are reluctant or feel guilty if they were not to participate (Haimes and Whong-Bar 2004:67).

This typology and other empirical work reveals a diversity of motivations when one is considering becoming involved in research. These range from altruistic motivations to expectations of personal or familial benefit, previous experience of benefit from research, perceptions of low risk or requirement of little active participation, and peer and cultural pressure (Haddow et al. 2004:23).

Reactions to the potential for commercialization of biobanks as research resources in this climate can be understood as follows:

(1) a perceived incompatibility of commercial drivers with research resources intended to support a health service for the public good (Levitt and Weldon 2005:318);

(2) a perceived unfairness of private profit-making from freely donated material to a public resource, seen as exploitation through commodification with a feeling of disrespect caused by commercial use (Haddow et al. 2006:278);

(3) the view that research data and results should be publicly owned, used for the public good, and made freely available (Haddow et al. 2006:278);

(4) a reluctant acceptance that investment from profit-making pharmaceutical companies is necessary for innovation in new diagnostics and possible treatments (Haddow et al. 2006:278);

(5) a willingness to accept commercial access if there is strict assurance that access controls would be in place;

(6) a need to build models for benefit sharing back to the community;

(7) a need for transparency and openness in relation to commercial interests as pivotal in earning and keeping trust (Sumner 2007).

This variety of motivations has yet to be considered systematically in current frameworks, although increased efforts emphasize their importance for the dynamics and the very possibility of research. Their further analysis is essential for better understanding of participants' aspirations, expectations, and values in the purpose and scope of research in order to build mechanisms for effective governance.

5.5 **Assumptions and ambiguities**

The purpose of this chapter is to emphasize that, in order to address these issues, it is important to clarify the conceptual basis for governing prospective biobanking research. In existing frameworks, the logic of disempowerment is pervasive and is further facilitated by enduring yet incongruous assumptions of altruistic surrender in research. These are exemplified, for example, in existing UK guidance on the use of human biological samples in research that 'donated material' is to be construed as a 'gift' involving a free and voluntary transfer with no expectation of return (Nuffield Council on Bioethics 1995:68; MRC Working Group on Human Tissue and Biological Samples 2001:para 2.2).

This framework imports a policy model that sees the transfer of samples solely as a question of 'altruism'. I define altruism as the 'action in the interests of another or the disposition to act in the interests of another' (Birks 2000:14). This approach has several shortcomings in that it erroneously imports the blood donation paradigm to the realm of research with human biological materials where the same circumstances do not apply, it relies on the false premise that potential research participants act exclusively out of charitable inclination and denies other interests that they may have, and ultimately it becomes a dangerous rhetoric that invites exploitation (Dickenson 2004:118).

This approach further introduces a significant, and unaddressed, paradox because in law a gift presupposes an underlying property right. One can only give as a gift what one owns (Laurie 2002:88; see also Chapter 9, section 9.4.2). When consulting existing guidance, one cannot help but wonder that if there can be no ownership of human tissue (when compared with the European Convention on Human Rights and Biomedicine position that the human body shall not as such give rise to financial gain), how can one transfer it to someone else as a gift? To the extent that existing guidance stipulates that 'any proprietary rights that the donor might have in their tissue would be transferred with the control over the use of the tissue to the recipient of the gift' (MRC Working Group on Human and Biological Samples 2001:para 9.2) and that the donor 'surrenders all interests' (Nuffield Council on Bioethics 1995:68), the question is left open. The current legal position implies that property interests exist, and this is awkward because when debating who has control over the transfer and use of research material, one needs to know what rights can be asserted, by whom, and to what extent.

By leaving this issue open or rather implied, the tension about whether participants can attain some degree of control in research is unaddressed in existing UK frameworks. As it currently stands, UK guidance does not resolve this ambiguity which can cause great unease towards participation, especially on a long-term basis. In social empirical research from Generation Scotland, the Ethical, Legal and Social Implications (ELSI) team considered the language of 'property' and 'ownership', used by the public as metaphors for 'control', as part of efforts to minimize the effects of private profit on recruitment (Haddow et al. 2006:272). Unambiguous solutions and mechanisms are needed, premised not necessarily on whether participants (who agree to give biological material for research purposes) can have property rights over samples but, mainly, on whether and how they can secure a better control over the *use* of samples and other rightful claims, for example for the time, effort, and resources they contribute to research, including claims for their return or destruction.

5.6 **Envisaging reciprocity**

A possible way of developing intuitive mechanisms that afford some degree of control to research participants throughout the research process would be to articulate and establish notions of reciprocity in research. In the last few years, scholars have begun to question the appeal to altruism with no expectation of return in biobanking so as to reconfigure the biobank paradigm as a model of 'exchange relationships', according to which some form of *reciprocal benefits* is expected (Haimes and Whong-Barr 2004:69).

Such a shift suggests the need to consider some form of benefits, however small or intangible, that could be offered to individuals or groups rather than focusing on educating participants not to expect any such benefits. Similar comments were made in consultations on the UK Biobank and its Ethics and Governance Framework in that the benefits of the biobank need to be more clearly spelled out, and that tangible benefits need to be reconsidered, as a way of enhancing the practicability and viability of a resource that requires high levels of long-term cooperation and trust (Sumner 2007).

There is a trend in current legal thinking to develop mechanisms for participants' control in research. This thinking would benefit further from a *reflective* understanding of research participation as a form of social *engagement*. In building reciprocal paradigms for research, reciprocity could be understood as a fundamental value that imposes obligations for mutual returns of favours and resources between participants, researchers, and society (Kanellopoulou 2009b). It would require an assessment of what appropriate returns would be, a commitment to mutual collaboration and support, and ways of monitoring such a collaboration over time.

In considering ways to redefine research relationships as reciprocal engagements that allow for a continuous return of favours between participants and researchers, a number of criteria can be developed for mutual consideration and agreement between researchers and participants on the following issues:

(1) a mutual understanding of what the *common good* that research biobanks promote is;

(2) practices that help the research *relationship* to exist as a dynamic engagement;

(3) mechanisms to evaluate the *contribution* of researchers and participants to the research process in equitable ways;

(4) an assessment of the *utility* of the research to all parties and to society as a whole;

(5) a *communication* and *negotiation* of parties' expectations in the research process;

(6) willingness to promote values of *fairness and justice*;

(7) awareness of and compliance with institutional constraints that may apply.

Such forms can be complemented by benefit-sharing mechanisms inspired by tenets of distributive justice. Recent proposals claim new approaches to take participants' interests seriously, but their analysis of is beyond the scope of this chapter. To the extent that such models warrant either limited or excessive control to research participants, further conceptual analysis is needed to develop workable notions of reciprocity in the evaluation of participants' contribution in research (Kanellopoulou 2008).

As part of such processes, new mechanisms could be developed for the representation and communication of research participants' interests on one hand, and for transparency and accountability on the other, based on their interests as well as on broader societal goals. One example is the proposal for the establishment of a Donor Approval Committee in the UK Biobank (Winickoff 2007), as part of a model that enables participants to have direct representation in determining how a collective public resource is allocated. Another example is the combination of material transfer agreements (MTAs) and EU standard contractual clauses in regulating cross-border flows of human tissue in an effort to develop and enforce clear safeguards for the interests of research participants in the transfer of tissue (see Chapter 13, section 13.5.3). Such proposals introduce meaningful ways for individual as well as collective *empowerment* that would enhance and support public trust and sustainable partnerships in research.

References

Appadurai, A. (1986) *The social life of things: commodities in cultural perspective*. Cambridge: Cambridge University Press.

Birks, P. (2000). The content of fiduciary obligation. *Israeli Law Review*, **34**, 3–38.

Busby, H. (2004). Blood donation for genetic research: what can we learn from donors' narratives. In R. Tutton and O. Corrigan (eds), *Genetic databases: socio-ethical issues in the collection and use of DNA*, pp. 39–56. London: Routledge.

Cambon-Thomsen, A. (2004). The social and ethical issues of post-genomic human biobanks. *Nature Review Genetics*, **5**, 866–73.

Cragg Ross Dawson (2000). *Public perceptions of the collection of human biological samples*. London: The Wellcome Trust.

Dickenson, D. (2004). Consent, commodification and benefit sharing. *Developing World Bioethics*, **4**, 109–24.

Generation Scotland (2009). *Generation Scotland website*. Available at: http://www.generationscotland.org/(accessed 26 November 2009)

Haddow, G., Cunningham-Burley, S., Bruce, A., and Parry, S.(2004). *Generation Scotland: public and stakeholder views from focus groups and interviews*. Innogen Working Paper, No. 20. Edinburgh: Innogen.

Haddow, G., Laurie, G., Cunningham-Burley, S., and Hunter, K. (2006). Tackling community concerns about commercialization and genetic research: a modest interdisciplinary proposal. *Social Science & Medicine*, **64**, 272–82.

Haimes, E. and Whong-Barr, M. (2004). Levels and styles of participation in genetic databases: a case study of the North Cumbria Community Genetics Project. In R. Tutton and O. Corrigan (eds), *Genetic databases: socio-ethical issues in the collection and use of DNA*, pp. 57–69. London: Routledge.

HGC (Human Genetic Commission) (2000). *Whose Hands On Your Genes?* London: HGC.

HGC (Human Genetic Commission) (2002). *Inside Information*. London: HGC.

HUGO (Human Genome Organisation) Ethics Committee (2000). *Statement on benefit sharing*. Available at: http://www.hugo-international.org/img/benefit_sharing_2000.pdf (accessed 5 January 2010).

Kanellopoulou, N.K. (2008). *Group rights in biolaw: a model approach*. Edinburgh: University of Edinburgh.

Kanellopoulou, N.K. (2009a). Advocacy groups as research organisations: novel approaches in research governance. In C. Lyall, T. Papaioannou, and J. Smith (eds), *The limits to governance*, pp. 193–215. Farnham: Ashgate.

Kanellopoulou, N.K. (2009b). Reconsidering altruism, introducing reciprocity and empowerment in the governance of biobanks in the UK. In J. Kaye and M. Stranger (eds), *Principles and practice in biobank governance*, pp. 33–52. Farnham: Ashgate.

Laurie, G.T. (2002). *Genetic privacy: a challenge to medico-legal norms*. Cambridge: Cambridge University Press.

Levitt, M. and Weldon, S. (2005). A well placed trust?: Public perceptions of the governance of DNA databases. *Critical Public Health*, **15**, 311–21.

MRC (Medical Research Council) Working Group on Human Tissue and Biological Samples (2001). *Human tissue and biological samples for use in research*. London: MRC.

Nightingale, P. and Martin, P. (2004). The myth of the biotech revolution. *Trends in Biotechnology*, **22**, 564–9.

Nuffield Council on Bioethics (1995). *Human tissue: ethical and legal issues*. London: Nuffield Council in Bioethics.

O'Neill, O. (2002). *Autonomy and trust in bioethics*. Cambridge: Cambridge University Press.

Office of Science and Technology and Wellcome Trust (2000). *Science and the public: a review of science communication and public attitudes to science in Britain*. London: Office of Science and Technology.

Petersen, A. (2007). Biobanks' 'engagements': engendering trust or engineering consent? *Genomics, Society and Policy*, **4**, 31–43.

Rothstein, M.A. and Griffin Epps, P. (2001). Ethical and legal implications of pharmacogenomics. *Nature Reviews Genetics*, **2**, 228–31.

Shickle, D. et al. (2003). *Public attitudes to participating in UK Biobank: a public consultation on issues relating to feedback, consent, withdrawal and access*. Research Report. Sheffield.

Smith, B.H., Campbell, H., Blackwood, D., et al. (2006). Generation Scotland: the Scottish Family Health Study; a new resource for researching genes and heritability. *BioMed Central Medical Genetics*, **7**, 74–82.

Sumner, J. (2007). *Public attitudes to biobanks and related ethics and governance issues*. Edinburgh: UK Ethics and Governance Council.

Terry, S.F., Terry, P.F., Rauen, K.A., Uitto, J., and Bercovich, L.G. (2007). Advocacy groups as research organizations: the PXE International example. *Nature Reviews Genetics*, **8**, 157–64.

Tutton, R. (2004). Person, property and gift: exploring languages of tissue donation to biomedical research. In R. Tutton and O. Corrigan (eds), *Genetic databases: socio-ethical issues in the collection and use of DNA*, pp. 19–38, London: Routledge.

UK Biobank. *UK Biobank website*. Available at http://www.ukbiobank.ac.uk/(accessed 26 November 2009)

Waldby, C. and Mitchell, R. (2006). *Tissue economies—blood, organs and cell lines in late capitalism*. Durham, NC: Duke University Press.

Winickoff, D.E. (2007). Partnership in U.K. Biobank: a third way for genomic property? *Journal of Law Medicine and Ethics*, **35**, 440–56.

Chapter 6

Taking solidarity seriously: do biobank institutions have a moral obligation to inform their patients about incidental health findings?

Christian Lenk

6.1 Introduction

It is a principle of research ethics, which also found expression in articles 14 and 33 of the Declaration of Helsinki[1], that every patient who participates in a research study should eventually have access to the newly developed therapeutic and diagnostic measures. However, in the case of medical research in the context of biobank institutions it is somewhat unclear what kind of benefit donors of tissue or body material might have. One obvious possibility in which most participants of such studies would probably be interested is the feedback of important incidental health findings (Bovenberg *et al.* 2009). It may be possible to define specific genetic findings which are relevant to individual therapy, which should be communicated to the donors (in the case of an epidemiological biobank) or the patients (in the case of a disease-related biobank). For example, when in the course of genetic screening tumour markers are found which show that the donor has cancer, he/she should be told to consider starting appropriate therapy. Nonetheless, it should also be accepted that the tissue donor may not wish to be informed of relevant findings on the basis of his general right not to be informed. Such relevant information should be defined *before* the project starts in order to ensure that the only information which is fed back to the donors is of high therapeutic relevance and not merely representative of a genetic probability. It is necessary that the information procedure accords with existing guidelines for patient information (for example, in the case of genetic diagnosis). The information should be communicated by the treating medical consultant (in the case of disease-related biobanks) or by the general practitioner (in the case of epidemiological biobanks). The strongest argument in favour of individual feedback to participants may be that participants could allege that the biobank's failure to pass on patient information resulted directly in damage to the donor's health. This chapter compares legal, ethical, and institutional guidelines from several European countries and discusses the

[1] World Medical Association (2008): '(14) The design and performance of each research study involving human subjects must be clearly described in a research protocol...The protocol should describe arrangements for post-study access by study subjects to interventions identified as beneficial in the study or access to other appropriate care or benefits...(33) At the conclusion of the study, patients entered into the study are entitled to be informed about the outcome of the study and to share any benefits that result from it, for example, access to interventions identified as beneficial in the study or to other appropriate care or benefits.'

different approaches in the framework of solidarity, which is often cited as a fundamental justification for undertaking research with population biobanks.

6.2 Why should anybody take part in a biobank research project?

Biobank projects have a problem in motivating potential donors to participate in their project. In particular, epidemiological biobank projects, which do not focus on research on a specific disease, have to make clear what kind of research enterprise patients' or test persons' body material is required for. The general question which participants in biobank research may pose is: What is my specific benefit when I donate body material? Also, what is the overall benefit when I participate in this study? Therefore most biobanks which have a publicly accessible website provide some explanations as to why patients or test persons should participate in their project (for people's actual reasons for participation, see Chapter 5, section 5.4.2). A review and comparison of the websites of some different biobank projects reveal three different lines of reasoning why research with biobanks is valuable.

The first argument uses a literal interpretation of the value of body material, although the definitions oscillate between pecuniary value and scientific value (see Chapter 15, section 15.5, for a distinction between different concepts of value). For example, the UK Biobank website contains a section entitled 'Why is it important that I take part?' and express the idea of value in the following way:

> UK Biobank is building a national treasure trove of health information, of an impressive scale, to be used by scientists in the future. (UK Biobank 2009a)

The Generation Scotland website is similarly ambiguous with its slogan: 'Addressing the Health and Wealth of Scotland' (Generation Scotland 2009).

The Sweden Medical Biobank website puts it the following way:

> The exploitation of VIP [the Västerbotten Intervention Project] is only in its infancy, but already several important scientific papers and doctoral theses, for example, on the role of viral infections in the development of cervical cancer, hormonal aetiology to breast and prostate cancer, and cardiovascular disease etc. [have been published]...Hopefully future funding will allow for additional staff positions for epidemiological researchers making use of the Medical Biobank as a goldmine for research on the causes of disease.' (Västerbotten Intervention Project 2009)[2]

A (critical) BBC article on the UK Biobank also used the gold metaphor and phrased it thus:

> It's hoped that the information collected will be the research equivalent of gold dust—containing the secrets of how genes and environmental factors conspire to make us ill. (Ghosh 2003)

Whether consciously or subconsciously, the term 'biobank' seems to inspire an astonishing number of popular science writers to compare (or equate?) scientific value with monetary value, despite the apparently routine understanding that these are two totally different 'currencies'. The message for the patients or research participants here is simply ambivalent. It might be 'Please help us to build a valuable source for research and to increase the "wealth of nations" (in the broader sense of the term)'. But it could also be interpreted as 'Please help us to explore the goldmine which is in your body, tissue, and cells'. However, these metaphors may also be interesting in the context of current public research policy, which presumes an adequate financial return for investment of public money. Unfortunately, investment in scientific enterprise is inherently risky and there is absolutely no guarantee that a possible scientific value will also be transformed in the

---

[2] The Swedish discourse on participation in biobank research is analysed in Chapter 7.

'true' (at least from the point of view of research politicians) or market value (Kattel 2008; Wade 2009).

The second line of argument is related to the notion of risk (for a detailed discussion of risk in the context of human tissue research, see Chapter 4), the fear of genetic diseases, and uncertainty as to whether somebody could have a predisposition for a specific disease. For example, Generation Scotland's website argues the following way:

> It's in the family. My auntie had it and so did my granddad. Now my brother's gone down with it. What about me and my children?
>
> Cancer, heart disease, stroke, mental health problems and many other common conditions tend to cluster in families. It's all to do with the genes we inherit, the environment we live in and our lifestyle. (Generation Scotland 2009)

Probably everybody (certainly individuals with a predisposition to hypochondria) knows the vague feeling that something might be wrong with their body, brain, or (even worse) genes, and one day a gruesome disease will break out and affect themselves and their own children. At the same time, one can also see a decisive change of tack in relation to classic genetics: it is not only the genes, but also the individual environment and lifestyle which can result in the onset of a genetically predisposed disease. It is precisely this phenomenon that sets the goals for biobanks: to bring these factors together and draw some conclusions regarding individual genetic risk. In this case the conclusions reached by patients and participants differ from those reached when following the first line of reasoning: it is not the scientific or monetary value which is mentioned here, but the often ambiguous knowledge which is produced by genetic analysis. Potential participants who take the paragraph cited above seriously would expect to receive some kind of information about potential familial health risks, but would probably also consider whether this kind of knowledge really is a benefit or whether it is some kind of burden for themselves and their families. It is far from clear whether population biobanks are prepared and able to meet the kind of expectations which can plausibly be derived from their own promises. This uncertainty holds for both general scientific findings derived from experiments undertaken in relation to whole cohorts and incidental findings of relevance to individuals or their families.[3]

The third line of reasoning has possibly the strongest normative implications in the ethical sense of the word, i.e. the altruistic appeal to participate in a biobank project for the well-being of others. For example, the first page of the UK Biobank website reads as follows:

> Have you received your letter inviting you to participate in UK Biobank? Do you want to do something good today? Not just for yourself, but for future generations? Signing up to take part in UK Biobank's unique research project is easy to do. (UK Biobank 2009b)

And further in the section 'Why is it important that I take part?':

> While the people to benefit most will be our children and their children, many of us may see our own lives touched by this exciting, innovative and far reaching research project. Please consider taking part in UK Biobank—and help us unlock the secrets of disease that will bring about a better life for all.

The perception of biobanks is placed on a societal level here; they are not seen as mere scientific projects but as helpful tools for society. The arguments construct a connection between the individual, future generations, and diseased persons. One could say that the implicitly existing genetic association between healthy and diseased individuals in an existing society is transformed into a

[3] This doubt is reinforced by Kanellopoulou's observation that in the current discourse on human tissue research the disempowerment of donors and patients is pervasive (see Chapter 5, section 5.5).

moral appeal for cooperation and assistance to enable a better society in the future. Compared with other medical projects, it is not the individual and his/her responsibility which are the focus of attention, but the individual as part of the collective. To take part or not to take part is not simply an individual choice, but has a moral relevance for the health and well-being of future generations. Therefore it is a language of altruism and solidarity which is chosen by the researchers in this context to motivate potential participants to join the biobank project. The potential participants are addressed as persons who have the choice (and maybe the duty) to help others. The reason why this is remarkable is the fact that this is rather unusual in a modern liberal society in which duties of this kind for its citizens are reduced to an absolute minimum (compared with earlier types of society).

The US Biobank Central website takes a comparable approach to the argument. In the section 'Why should a patient donate?', one can read the following:

> The intent of biospecimen research is to apply genetic science to human health in order to improve healthcare. Biospecimen research is inherently dependent upon an individual's willingness to donate biological materials such as DNA, blood, or tissue fragments to the research community for altruistic purposes. Not all of the tissue taken for a biopsy is required for diagnosis, so researchers can use the 'extra' tissue to improve understanding about a disease, develop targeted therapies, and enhance quality of life for affected individuals.
>
> Many patient donors report that they wish that 'something good' will result from their disease, and hope that by donating, they will contribute positively to finding cures for others. Since many diseases have a genetic aspect, many patients believe that they will in this way be helping family members as well as society in general. (Biobank Central 2009)

In this text, the donors and patients themselves are imagined as persons with an altruistic self-understanding. This fits in with the interpretation of the transfer of body material as a 'donation', i.e. an act which is normally perceived as unselfish and altruistic. The addressees for altruistic behaviour in this text are potential family members or the society at large. The unconcerned individual makes a voluntary genetic contribution to the benefit of his co-citizens and society. However, the motives of the biobank and the researchers in the background remain obscure. Are their activities also to be seen in the context of solidarity? Or is it primarily the donors and patients who are obliged to show solidarity?

6.3 **The meaning of solidarity in the context of biomedical and genetic research**

In the context of this chapter, the term 'solidarity' will be defined as 'a voluntary, reciprocal, positive, group-specific and protective moral obligation' which can also be transformed into contractual and institutional obligations. The motivation to act in the framework of solidarity is normally based on a free and voluntary decision. Solidarity as such can only rarely be enforced, although it is often demanded by others.[4] Because it is group-specific and those who act or should act according to the framework of solidarity are normally members of this group, solidarity cannot simply be described merely as unselfish behaviour. This is because such actions bring benefit to the group, and as such—because the acting person is part of this group—also to those who act under the banner of solidarity. According to the conceptual distinctions of philosophy,

---

[4] To be enforceable, the duties of solidarity have to be transformed by political decision into legal duties, for example to become an obligatory member of the public health insurance scheme and to pay regular taxes or fees for health services.

solidarity is not a negative duty (to avoid a certain behaviour, for example damaging others), but a positive ethical duty (to help others). Solidarity is concerned with the protection of others who face serious problems in such areas as human or social rights or in the fields of health and work.

Starting from such a definition it is prudent to ask what the relationship between solidarity and biomedical (and especially genetic) research might be.

The history of medicine provides evidence of several kinds of argument known to support mechanisms of solidarity in connection with genetic findings. Classic eugenics demanded a certain kind of solidarity of potential parents with future generations in particular and the whole *Volk* (peoples) as a biologically defined group in general. Any person with problematic genetic or phenotypical features, according to the determination of genetics at that time, was obliged to restrict his/her reproductive behaviour (which was indeed formulated as a kind of ethical duty). A considerable number of states (see, for example, Nazi Germany's Law for the Prevention of Hereditary Defective Offspring) also enshrined compulsory sterilization in their legal systems as an expression of the prevalent opinion that sterilization would be a minor sacrifice for the individual compared with the genetic advantage to the whole population.

A less extreme example is the idea of solidarity for the financing of health-care systems, which is based on the idea that there should be active solidarity with sick individuals, especially in cases where these individuals have no responsibility for the onset of their disease. Of course, this is also the case with genetic diseases. Health problems are clearly seen as a category of personal difficulties outside or at the margin of personal responsibility (a description which is appropriate in the majority of cases but certainly not for all kinds of diseases). In contrast with other problems of the individual, which one may be required to overcome alone, difficult health conditions are perceived as a category of problems which entitles the individual to the support of society.

In contrast with these two connections between solidarity and genetics, the motivation of patients to take part in research projects is often seen not only as the potential benefit for the participants themselves, but also as the potential benefit to the group of patients with the same health condition. This situation is also privileged by principles of research ethics and international instruments in the area because, in these cases, non-competent patients can be included in research without explicit informed consent (on the basis that there is only a minimal risk of damage and harm).[5] In such cases, it may be true to say that the demand for solidarity facilitates the conduct of biomedical research.

From this point of view, patients with a genetically predetermined disease are a special group defined by common genetic characteristics and a shared and comparable 'genetic destiny'. Past experiences with genetic research and biobanks show that such patient groups tend to organize themselves and also to develop joint activities to foster research regarding their condition and to protect the rights of their group by ethical and legal action (see Chapter 5, section 5.3).

One example for this type of cooperation is the Greenberg case. The case dates back to an initiative of parents of paediatric patients who suffered from Canavan disease. The initiative led first to the establishment of a patient registry and later to the first gene test for the disease. Canavan disease is a hereditary and autosomal recessive degeneration of the central nervous system and manifests in early childhood. In 1987, Daniel Greenberg, the father of two children who suffered

[5] *Declaration of Helsinki* 2008: article 27: 'For a potential research subject who is incompetent, the physician must seek informed consent from the legally authorized representative. These individuals must not be included in a research study that has no likelihood of benefit for them *unless it is intended to promote the health of the population represented by the potential subject*, the research cannot instead be performed with competent persons, and the research entails only minimal risk and minimal burden.' (emphasis added). See also Article 17 of the *European Convention on Human Rights and Biomedicine.*

from Canavan disease, tried to find a scientific institution which would cooperate in the search for the genetic cause of his children's disease. In the course of this search, he contacted the physician and scientist Reuben Matalon, who agreed to work together with other families who were affected by Canavan disease. As well as establishing the registry, the families also supported the researcher with body material from their children and a large amount of money for financing of the research. In 1993, Matalon successfully identified the relevant gene. This was the precondition for the development of a genetic test for Canavan disease which, by 1996, enabled the Canavan Foundation to give potentially concerned persons free access to genetic testing.[6]

Another example of the cooperation of patients and research institutions is the Eurobiobank project for genetic research in the context of rare diseases. This project is a product of the initiative and, in part, the activity of patient groups. The foundation of the Eurobiobank research consortium was initiated by letters from families and patients who were affected by rare diseases and wanted to participate in genetic research by giving blood or tissue samples. Currently Eurobiobank comprises 170,000 samples of rare diseases and is composed of 15 partners from seven European countries. In the case of rare diseases, combination of different collections is of exceptional importance in order to enable genetic research to be performed on a sufficiently large number of samples.[7]

It can be concluded from the experiences outlined above that the solidarity shown by patients can lead to remarkable efforts in biomedical and genetic research and decisively improve the circumstances for medical research. It also follows that the discussion of 'solidarity' in the context of research projects should not be taken as mere rhetoric—it clearly has some substance. However, the interesting question seems to be whether the situation as described above fits into all cases of biobanks or whether there are situations which do not fit into the framework of solidarity.

It is reasonable to assume that the bonds of solidarity will be experienced in a stronger way the smaller and better defined the respective patient group and the more serious the situation experienced by the members of this group. When we compare this description with the prevalent forms of biobank organizations, we find that there are huge differences in relation to solidarity between the two main biobank types—epidemiological and disease-related. In the case of epidemiological biobanks, such as UK Biobank, tissue and blood samples originate from the 'normal' population with a corresponding normal distribution of health and disease. In this case, there is a massive group of donors with very diverse diseases (or no disease at all) which makes it rather unlikely that the donors experience any stronger feeling of solidarity. In the second case, the disease-related biobanks, the situation is quite different. Take, for example, children with congenital heart failure, HIV patients, or patients with a genetic disposition for Alzheimer's disease. In these groups, it is very plausible that the donors have a shared (genetic or other) 'destiny' and a strong motivation to act in accordance with the framework of solidarity.

We can call such a phenomenon, which results in the self-organized action of patient groups, 'spontaneous' solidarity. In contrast with these spontaneous manifestations of group solidarity, there are also appeals for an extension of solidarity in the normative framework of society in the way shown by the preceding quotations from the UK Biobank and the US Biobank Central show. Additionally, there are theoretical considerations such as those expressed by in a 2001 article by Chadwick and Berg (2001) in an article on solidarity and equity. These authors query whether

---

[6] Although this undertaking started out with solidarity in mind, later on there was severe conflict between the Canavan Foundation and the research institution; for further information see Marshall 2000:1062; *Greenberg* v. *Miami Children's Hospital Research Institution* 264 F.Supp. 2d 1064 (2003; US District Court, S.D. Florida, Miami Division).

[7] For further information, see www.eurobiobank.org

society should extend the notion of solidarity to a kind of duty to give samples for research purposes.

> In recent years, there has been a very strong emphasis on the individual's rights, for example, to refuse to participate in genetic research or to refuse to have health information recorded in medical or genetic databases. It is considered a right of people participating in medical research to withdraw from a study at any time and to demand that one's sample is given back, regardless of the damage to research or researchers. It is not obvious, however, why a right to refuse to participate in genetic research, when it could be to the benefit of others, should be overriding. On the contrary, it could be argued that one has a duty to facilitate research progress and to provide knowledge that could be crucial to the health of others. This principle of solidarity would strongly contradict a view that no research should be conducted if it would not directly benefit those participating in a study. (Chadwick and Berg 2001:320)

However, such an argument sets up a contradiction between the principles of solidarity and autonomy and aims to extend the notion of solidarity *beyond* the limits of voluntary participation in order to come to a kind of duty for genetic examination for the sake of research. This would be a strong breach of the long-standing tradition of voluntary research in an area where the self-determination of the individual appears paramount to avoid the misuse of genetic data.

6.4 **Solidarity: mutual or one-sided ethical duty?**

It appears to be undecided whether biobank research builds on a *mutual* relationship such as solidarity or a one-sided demand to the tissue donors such as altruism. Comparing the projects mentioned in the previous section, it is doubtful whether a biobank project with a very heterogeneous group of donors, which consists of healthy people as well as persons with very different diseases, satisfies the criteria of solidarity in a sufficient fashion. There can be a broad kind of solidarity with people who have a disease, in particular a genetic disease, but this seems to be different in the case of the patients with Canavan disease and other rare diseases cited above.

Another important question is whether biobanks should also be seen as institutions which, because of their reference to the framework of solidarity, are obliged to act towards their donors out of solidarity. Although most biobank projects have no direct therapeutic potential, one important issue is the question of whether there is an obligation to inform patients about individually important health findings. Some national recommendations, for example from Austria (Bioethikkommission beim Bundeskanzleramt 2007:41), Germany (Nationaler Ethikrat 2004:68), and Switzerland (SAMW 2006:9), demand such an information policy for findings which are essential to life. For example, the Swiss Academy of the Medical Sciences (SAMW) argues:

> Donors have the right to be informed about findings of diagnostic or therapeutic relevance (right to know). This does not apply to irreversibly anonymized samples and data. In general, the information will be given by the responsible physician. He is to ensure adequate counselling. The biobank management secures the flow of information. The donor can relinquish the right to later information (right not to know). (SAMW 2006:9, §4.7; translation by the author)

As an illustration, UK Biobank expressly excludes the communication of individual findings to donors. This form of regulation by UK Biobank was fundamentally criticized by Johnston and Kaye (2004) from both an ethical and a legal perspective, arguing from the point of view of the English legal system. One of their arguments touches the appropriateness of the decision not to inform patients—whatever health finding there may be during the work of the UK Biobank—in relation to the demands on the side of the patients:

> It is also worth recalling that the participant will not be paid for taking part, will be asked to supply confidential information over a period of up to twenty years, undergo a health screen and continue to

> update the project with relevant information. This altruistic action suggests that in cases of serious treatable conditions it would be unfair, unjust and unreasonable not to impose a duty to provide individualized feedback. (Johnston and Kaye 2004:253)

This means that a point is reached where, through the density of interaction, the relationship between the patient and the biobank becomes quite close, and it becomes more and more difficult from an ethical and legal point of view to withhold the information from the patient. However, there is good evidence that other jurisdictions assess the question of the necessity of patient information more seriously than in the English context. The German Federal Supreme Court of Justice explicitly equates the withholding of important information with medical negligence:

> It is a serious medical mistake in the treatment, when the patient will not be informed about a threatening result, which gives grounds to immediate and extensive measures, and when the necessary medical treatment is refused to him. (Franz and Hansen 1997:93; translation by the author)

This argument clearly has to be seen in the context of a presumed duty of care of the physician in relation to his patient. When the physician knows about a threatening diagnostic information, he/she is responsible for avoiding the potential harm which can be caused by a disease process to the detriment of the patient. Otherwise, although the disease process is in no way induced by the physician, he carries the responsibility for this avoidable harm. In such circumstances, the question of liability for damage to the patient arises. It follows that anybody who stands in a professional relation to others carries a specific responsibility to avoid unnecessary harm which occurs, for example, from exclusive knowledge that he/she possesses.[8]

6.5 **Conclusions**

The definition of solidarity given here permits the explanation of phenomena of social cooperation among patient groups and among patients, researchers, and research institutions. However, there are also situations in the field of research with human tissue and biobanking, such as large heterogeneous donor groups, where acting out of solidarity seems to be rather unlikely. From the point of view of medical ethics, the reference to biobank research in the framework of solidarity is reasonable when two conditions are satisfied:

1. The donor or patient group should share the comparable experience of a similar disease entity. This must not mean that only the concerned persons themselves can show solidarity, but may well mean that relatives and families can be included in this framework of solidarity.
2. Solidarity in the narrow sense of the word means a mutual relationship, but not the one-sided obligation of the donor or the patient to give his/her sample. It follows that biobank institutions should develop and foster methods and ways of showing solidarity towards their donors or patients, for example the provision of feedback of information of therapeutic relevance).

There may also be additional possibilities for combining the framework of solidarity with other biobank projects (e.g. epidemiological biobanks). A large epidemiological institution such as UK Biobank could support specific patient groups and offer its donors contact to these groups in cases of identification of the respective disease. The epidemiological biobank would then be a combination of different disease-related biobank projects rather than one biobank with a very

[8] A similar argument is given by Johnston and Kaye (2004:242) for the ethical guidelines of the UK Biobank: 'If findings did come out of the research of the UK Biobank that did point to a serious, treatable condition, it would be very hard to sustain a moral argument that this information should not be fed back to identifiable affected individuals'.

heterogeneous group of donors. This could be an important step not only towards an understanding of biobanks as institutions of preventive medicine, but also towards responsible care for potential and real patients.

References

Biobank Central (2009). *Why should a patient donate?* Available at: www.biobankcentral.org/translational/donate.php (accessed 14 December 2009).

Bioethikkommission beim Bundeskanzleramt (2007). *Biobanken für die medizinische Forschung (Biobanks for medical research). Report of the Bioethics Commission at the Federal Chancellor's Office.* Vienna: Bioethics Commission.

Bovenberg, J., Meulenkamp, T., Smets, E., and Gevers, S. (2009). Biobank research: reporting results to individual participants. *European Journal of Health Law*, **16**, 229–47.

Chadwick, R. and Berg, K. (2001). Solidarity and equity: new ethical frameworks for genetic databases. *Nature Review Genetics*, **2**, 318–21.

Franz, K. and Hansen, K.-J. (1997). Aufklärungspflicht aus ärztlicher und juristischer Sicht (The duty of information from a medical and legal point of view). Munich: Hans Marseille Verlag.

Generation Scotland (2009). *Generation Scotland website.* Available at: http://www.generationscotland.org/(accessed 14 December 2009).

Ghosh, P. (2003). *Will biobank pay off?* BBC News Online. Available at: http://news.bbc.co.uk/2/hi/health/3134622.stm (accessed 14 December 2009).

Johnston, C. and Kaye, J. (2004). Does the UK Biobank have a legal obligation to feedback indivdual findings to participants? *Medical Law Review*, **12**, 239–67.

Kattel, R. and Suurna, M. (2008). The rise and fall of the Estonian Genome Project. *Studies in Ethics, Law, and Technology*, **2**, 1–22.

Marshall, E. (2000). Families sue hospital scientist for control of Canavan gene. *Science*, **290**, 1062.

Nationaler Ethikrat (2004). *Biobanken für die Forschung (Biobanks for Research).* Berlin: Nationaler Ethikrat.

SAMW (Schweizerische Akademie der Medizinischen Wissenschaften) (2006). *Biobanken: Gewinnung, Aufbewahrung und Nutzung von menschlichem biologischem Material. Medizinisch-ethische Richtlinien und Empfehlungen (Biobanks: procurement, storage and usage of human biological material. Medico-ethical guidelines and recommendations).* Basel: SAMW.

UK Biobank (2009a). *Why is it important that I take part?* Available at: www.ukbiobank.ac.uk/assessment/takepart.php (accessed 14 December 2009).

UK Biobank (2009b). *Improving the health of future generations.* Available at: http://www.ukbiobank.ac.uk/(accessed 14 December 2009).

Västerbotten Intervention Project (2009). *Samples from 100,000 unique individuals in the freezers!* Available at: www.umu.se/phmed/naringsforskning/Medical_Biobank/indexB.htm (accessed 15 October 2009).

Wade, N. (2009). A genetics company fails, its research too complex. *The New York Times*, New York edition, B2 (18 November 2009).

World Medical Association (2008). *Declaration of Helsinki. Ethical principles for medical research involving human subjects.* Available at: http://www.wma.net/en/30publications/10policies/b3/index.html (accessed 22 December 2009).

Chapter 7

Beyond the dichotomy of individualism and solidarity: participation in biobank research in Sweden and Norway

Katharina Beier

7.1 Introduction

Biobanks collecting human tissues together with data on lifestyle, environmental exposures, and genetic materials are of increasing importance in medical therapy and research. Owing to unquestioned sample storage routines in pathology and the perception of bodily leftovers as waste materials, the contribution of donors to the establishment of biobanks has often been neglected. At present, however, new fields and methods for the processing of human tissue samples have not only sparked the interests of researchers, industry, and law enforcement agencies, but also shed new light on the role of donors in this context. In particular, the research community is 'coming to recognize an increasingly broader range of subject's interests relevant to their research participation' (Wendler 2002:45), yet there is far less agreement on the conceptual framework for participation in biobank research. This chapter takes up the issue by drawing particular attention to Sweden and Norway. This focus is due, first, to the fact that both countries displayed an early awareness of the impositions confronting participants in biobank research. As early as 2003, the Swedish *Riksdag* and the Norwegian *Stortinget* (the parliaments of the two countries) passed acts on biobanks that also comprise *research* in human biological samples.[1] Secondly, as Scandinavia has a long tradition of epidemiological research which started in the first half of the twentieth century, Swedes and Norwegians are expected to be familiar with this type of research. This is emphasized by the fact that the Nordic countries host 'disproportionally large numbers of collected human samples' (Nobel 2008:12).[2]

The ethical debate on participation in biobank research is divided between communitarian approaches that stress the importance of solidary contributions to the common good (Wilson 2004:81; Knoppers and Chadwick 2005:76) and liberal approaches that point to personal benefits such as tailor-made medication that fits the needs of the individual patient. This contrast can be explained by the nature of biobank research in particular as well as the character of modern pluralistic societies in general: whilst the value of biobanks results from the *collective* nature of the collected samples and data, the success of biobank research depends upon an *individual* readiness to donate one's samples to research. However, this individual contribution cannot be taken for

[1] This is particularly noteworthy because the European Directive 2004/23/EC is solely concerned with therapeutic applications of human tissues.

[2] According to figures from the Public Population Project in Genomics (P3G), out of 82 large-scale national research cohort studies with a total of 7.8 million subjects, 2 million come from the Scandinavian countries. This matches 25% of the subjects currently enrolled worldwide (Nobel 2008:13).

granted in modern liberal societies (potential incentives for participation are outlined in Chapter 5, section 5.4.2, and Chapter 6, section 6.2). Thus biobank research is left with the obligation to provide a justification for this personal engagement. Whilst the contributions to this debate have generally oscillated between individualistic and common good arguments, the following argument aims at transcending this dichotomy. First, in section 7.2 the legal frameworks adopted by the two countries are outlined by taking public discourse on biobank participation into account. It will be argued that in both countries neither an outright individualistic nor a fully developed solidarity approach is prevalent. Rather, a conjunction of both perspectives is observed: the perception of biobank research through the lens of the common good *and* the protection of individual interests. In order to provide an explanation of this feature, two concepts that promote the reconciliation of these perspectives are highlighted in section 7.3. The Swedish welfare state can be shown to support 'state individualism' (Berggren and Trägårdh 2006), which rests on a reciprocal relationship between the state and the individual citizen, and in quite a similar way the Norwegian *dugnad* tradition (Ursin and Solberg 2008) strengthens reciprocal assistance amongst its citizens. However, the balance of individual and communal interests is still sensitive. It depends upon people's trust in research which, amongst other factors, is due to transparent regulation and straightforward information regarding the actual objectives of biobanking. Finally, based on these findings, some general conclusions regarding the prerequisites for participation in biobank research will be drawn.

7.2 **Biobanking in Sweden and Norway**

Swedish biobanks emerged from clinical routines and quality assurance measures. Currently, about 60 million samples are stored (Ursin et al. 2008:179), equivalent to 'seven samples per living Swede' (Hoeyer 2004a:6). As of July 2007, 651 biobanks were registered with the National Board of Health and Welfare (Nobel 2008:9). In 2002, the Swedish government initiated a National Biobank Programme for promoting systematic biobank research. It is based on the participation of major biobanks, the most prominent of which is the Medical Biobank in Umeå. Established in 1985 in the region of Västerbotten, it promotes research in cardiovascular diseases and retains about 265,000 blood samples from around 156,000 individuals. An even larger database comprising 3.3 million samples is the PKU Register (named after the disease phenylketonuria) at the Huddinge Hospital. Since 1975, all newborns have been tested for this metabolic defect, as well as for three other severe diseases, immediately after birth. There is also a twin register at the Karolinska Institute in Stockholm. In 2007, Sweden launched a national biobanking project (LifeGene) which will recruit 500,000 citizens aged 0–45 from all over the country for prospective and comprehensive health studies. In addition to the collection of biological materials and individual health and lifestyle data, genetic analysis is also envisaged (https://www.lifegene.se).[3]

In Norway, systematic collections of biological materials can be traced back to 1926 (Jansen 2004:141). The country's most valuable collections stem from population cohort studies carried out in the 1980s and 1990s, some of which are still ongoing. BioHealth Norway is a large population-based research platform for the accomplishment of genetic epidemiological research.[4] It is based on two core cohorts: the Mother and Child Study (MoBa), currently comprising about 270,000 individuals, and the Cohort of Norway study (CONOR), consisting of eight different

[3] A pilot study with 5000 participants started in summer 2009; the full-scale roll-out is planned for 2010.

[4] It was established due to a grant from the Norwegian Functional Genomics Research Programme (FUGE) and is hosted by the Norwegian Institute of Public Health—a research collaboration of five Norwegian universities.

cohorts from various Norwegian districts. The largest cohort is the Nord-Trøndelag Health Study (HUNT) at the Norwegian University of Science and Technology (NTNU) in Trondheim. It consists of three consecutive studies covering the period 1984-2008 with more than 100,000 participants from the county of Nord-Trøndelag. After its completion, BioHealth Norway will comprise biological samples and health and exposure data from about 500,000 Norwegians of all ages,[5] equivalent to a tenth of the population.

7.2.1 The framework of biobank participation in Sweden

Although biobanks did not attract any special attention in Sweden for a long time, the medical authorities realized the need for improved protection of donors after commercial use of biobanks was revealed by the media (Nilsson and Rose 1999; Trägårdh and Ringman 1999). In 2003, the Biobanks in Medical Care Act (BMCA) came into force (Biobanks in Medical Care Act 2002), which stipulates one of the strictest consent provisions within Europe (Rynning 2003:119), possibly in the world (Hansson and Björkman 2006:285).[6] Donor consent is required not only for the use of their samples in research, but also for every sample taken in the course of health care, provided that the storage of samples exceeds 2 months (Socialstyrelsen 2002:Chapter 1, section 3). In particular, the individual's choice is restricted to three options: (1) storage may be rejected, which implies that samples have to be discarded or anonymized; (2) donors may restrict the usage of their samples to medical diagnosis and therapy; (2) they may allow their samples and data to be used for research and educational purposes.

For the BMCA's enforcement, Swedish county councils (*landstingar*) released a range of leaflets in which the *personal* health of the patient is emphasized as the main incentive for allowing samples to be stored. In particular, it is stated that biobanks are 'repositories of information—from you and for you. It is because of you, for your security that samples are taken, analysed and stored' (Landstingen 2004:2).[7] Since the BMCA's main concern appeared to be with the empowerment and personal benefits of the *individual* patient (see also Ursin et al. 2008:182), 13 Swedish researchers complained in an open letter that 'the traditional Swedish values of general solidarity with the care for weak and ill fellow human beings had to give way to a distinctive individualistic ideology' (Adami et al. 2002:10).

The interests of individual donors are also stressed by the Ethics Policy of the LifeGene project. For example, donors are able to access information regarding their personal health by obtaining 'a password to his/her personalized page on the LifeGene homepage' (LifeGene Ethics Group 2009:4) which reveals blood test results, blood sugar, cholesterol, blood pressure, pulse, weight, waist and hip measure, adiposity, hearing, and lung function (LifeGene Informationsbroschyr 2009). In this way, probands have at least some overview of their individual samples and data.

However, there are also features acting as a counterbalance to an outright individualistic framework. For example, whilst the procurement of samples requires explicit consent according to the BMCA, this is not necessarily true for related data (Personal Data Act 1998:section 10, 18). This provision bears on the EU Data Protection Directive's specification that 'member states may, for reasons of substantial public interest, lay down exemptions' regarding the protection of personal data (Data Protection Directive 1995:article 8, paragraph 4).[8] Concerning research and statistical

[5] The third study was only completed in 2008 and the results are still being evaluated.

[6] For a more detailed analysis of the Swedish legal framework for biobanking, see Beier 2009.

[7] All English translations of quotations originally in Swedish or Norwegian are by the author.

[8] This is facilitated by the fact that data protection in public health matters is explicitly not harmonized by the EU law.

purposes, section 19 (1) suggests a specific weighing of individual against collective interests: If the latter are sufficiently strong and individual risk is estimated to be low, consent may be waived. However, this makes 'almost all non-consensual processing of personal data in relation to biobanks and genetic databases lawful'(Helgason 2007:101). In fact, a breach of the code for personal data is acknowledged for those cases where samples alone will not be sufficient to elicit scientific value (Socialstyrelsen 2002:Chapter 6, section 1). Since the term 'scientific value' is not defined any further, it is questionable whether non-disclosure can be expected to be the default.[9]

The LifeGene project even employs the communal framework as an explicit strategy: 'Participation will be presented as an opportunity to contribute to a resource that may, in the long term, help enhance other people's health' (LifeGene Newsletter 2009:7). The enrolment of 500,000 participants is promoted as a 'national project', thereby appealing to patriotic and solidary sentiments: 'Swedish people will commit themselves to participating in LifeGene by giving samples' since they are generally 'interested in giving such contributions to society' (Carlsson 2008). On account of this, donors are expected to give their *general* consent to using their samples for research. Whilst this displays a deviation from the specific consent provisions of the BMCA, the Swedish research community favours its revision by arguing that 'an opt-out system, similar to the one being used in Denmark, should be considered' (Nobel 2008:9).

At the same time, the Swedish discourse on biobanking has taken pains to avoid the impression that the common good can only be achieved by individual sacrifice. Rather, the storage of samples is promoted as *quasi-naturally* serving the interests of others or the community in general. For example, although the BMCA restricts the acquisition of samples to the health-care context, this does not exclude their application in research. By perceiving research 'as an integral part of health care' (Landstingen 2005:6), the attending physician is expected to obtain consent for using tissue samples for *both* therapy and research (Socialdepartementet 2008:10). Thereby, personal and collective interests are presumed to form a continuum: 'If you are asked whether your samples may be stored, you should naturally think of your own benefit first. But you also have reason to think of others. To collect samples within a biobank is to the best of all' (Landstingen 2004:2). Potential conflicts between the individual and the requirements of solidarity are put aside by the optimistic assertion that 'it is primarily weak people who benefit from research...' (Gentekniknämnden 2004:17). Given that these promises are by no means self-evident in the context of biobank research, the suggested alignment of individual and collective interest calls for an explanation. However, before this is provided, an outline the Norwegian legislation and public perceptions in this field will be given.

7.2.2 The framework of biobank participation in Norway

Like the Swedish BMCA, the Norwegian Biobanks Act came into force in 2003. However, the enactment of the *Helseforskninsgslov* (Act on Medical and Health Research (AMHR) 2008) introduced some revisions. By renaming the initial act the Clinical Biobanks Act research biobanks are now covered by the AMHR (section 56), but research activities are clearly restricted to the generation of 'new knowledge about health and disease' (section 4). In addition, it is stated that 'the participants' welfare and integrity shall have priority over scientific and social interests' (section 5). For the enforcement of this maxim, informed, voluntary, and expressed consent is stipulated as main rule (section 13). In addition, 'in the event of substantial changes to the research project', the act prescribes additional consent (section 15).

[9] For example, in the recently started LifeGene project it is proposed that personal data from different registries will be linked using the ID number that each Swedish citizen is provided with.

However, the Norwegian legal framework also steps beyond a purely individualistic approach by stressing the community's interests in biobanking. In particular, the newly approved AMHR allows 'human biological material collected by the health service in connection with diagnosis and treatment' to be used without the individual's consent, provided that the regional committee for medical and health research ethics concludes that 'the research in question is of significant interest to society, and the participants' welfare and integrity are ensured' (section 28). The option of obtaining the donors' broad consent instead of their specific approval also alleviates restrictions on biobank research. Furthermore, in exceptional cases of 'compelling private or public interests', it is the right of the King to 'decide that human biological material may be surrendered to the prosecuting authorities or to a court of law' (section 27). Judged on these provisions, the AMHR provides considerable relaxation of the consent provisions of the initial Biobanks Act.

The importance of the community in motivating participation in biobank research is also stressed by the HUNT population studies. The HUNT biobank is seen as a 'national resource' that works for the good of all, and citizens take pride in contributing to it (NTNU 2004:42; Skolbekken et al. 2005:345). As in Sweden, biobank research in Norway builds on the expectation of a distinctive readiness for participation (Kvalvaag 2004). In fact, the HUNT studies have been blessed by high participation rates. In the Norwegian Public Report[10], this is explained by the fact that, although some 'may themselves have expectations that the disposition for a disease will be revealed (for example, high cholesterol levels), or merely want a free health check, others participate out of a notion of collective solidarity' (NOU 2005:41).

As well as the collective benefits of biobank-related research, the HUNT studies also point out to individual gains. Although direct economic benefits are excluded, participants may still receive some indirect rewards. For example, the second HUNT study builds on a close relationship between the donor and his practitioner, implying that 'all participants with clinical findings indicating pathology were advised to see their family doctor, who also received clinical results from the health study itself' (Holmen et al. 2003:19). Thus individuals will not only have the chance of getting a free health-check, but may also learn about dispositions they might otherwise not have known about. On the other hand, feedback on *genetic results* is excluded (Holmen et al. 2004) since 'the idea of "giving back" in biobank research should primarily be understood on a collective level' (NTNU 2004:26) (for a general discussion on whether there is a moral obligation for individual feedback, see Chapter 6). However, as this should not be to the detriment of 'individual-oriented services' it is suggested 'that a part of each individual's sample be reserved for health service use for the diagnostic or therapeutic benefit of the individual in question' (NTNU 2004:48).

In the Norwegian discourse individual and communal interests are not perceived as mutually exclusive: 'A closer look…shows that…in most situations, research participants, society, research, researchers and economy have overlapping interests' (NOU 2005:129). The creation of 'closer ties between the (potential) donor community and the research teams' is displayed as a crucial precondition 'to harmonize and balance the respect for individual rights, consent and privacy with the interests of the research community' (NTNU 2004:43). Although this harmonious picture might be a justification rather than a universally valid description of the constellation of interests in biobank research, the alignment of various stakeholder interests was successful in the establishment of HUNT Biosciences Ltd. Its creation in 2007 was driven by the insight that the maintenance costs might hinder its effective utilization. Since the 'commercial exploitation of research participants, human biological material and health information, as such' is excluded by the

[10] *Norges Offentlige Utredninger* (NOU) are reports published by a panel or committee that is appointed by the Norwegian government. The parliament may request the government to establish such a committee.

AMHR (section 8), HUNT Biosciences is publicly owned.[11] According to an agreement between NTNU and HUNT Biosciences, since 2009 private companies can access samples and data in the HUNT biobank for commercial purposes. However, access rests on three conditions that provide limitations to a fully commercial model. First, if a company wants to use samples with the intention of gaining economic benefit, the approval of the HUNT review board, the regional research ethics committee, and the data protection authority is required. Secondly, all enterprises and projects that obtain data from HUNT samples are obliged to return these within a stipulated period. Finally, even economic gains need to be re-deposited in the HUNT biobank to be re-invested in medical research and health improvement work. Thus the users of the biobank directly contribute to its scientific and economic value. Given these unusual features, HUNT Biosciences is at best described as a '*for-profit-for-public* company' (Solum Steinsbekk et al. 2009:148).

Having described how individual and collective incentives for participation in biobank research in Sweden and Norway have been aligned, an explanation of this feature is provided in the next section.

7.3 Explaining biobank participation in Sweden and Norway

Some authors perceive the donation of tissue per se as participation 'in a social activity' (Steinmann, 2009:282). This perception builds on the assumption that 'the individual's donation is embedded in a network of actions that are all driven by solidarity and cooperation towards the common good' (ibid.). Admittedly, in light of the Swedes' and Norwegians' empirically proven readiness to leave their samples to research (Hoeyer, 2004b; Nilstun and Hermerén, 2006; Johnsson et al. 2008; Holmen et al. 2004; FUGE nyhetsbrev, 2007) it is quite tempting to explain this feature by strong collective ties in the respective societies. In fact, both countries stand for the Scandinavian model of welfarism that is known for its solidarity commitments. For example, in Sweden, the metaphor of the 'people's home' (*folkhem*) is proverbial for an attitude marked by equality, solicitousness, cooperation, and a readiness to help others.[12] In Norway, the notion of solidarity is embodied by the *dugnad* tradition which refers to a voluntary work that is carried out by the members of a community for the good of all. However, a deeper analysis of these concepts' implications reveals that participation in biobank research cannot solely be explained by people's concerns for the common good but is also the result of individualistic ambitions. Hence, although the Swedish 'state individualism' and the Norwegian *dugnad* tradition are promising candidates for bridging the gap between communitarian approaches on the one hand and individualistic at the other, there are also some limits of this realignment in the field of biobank research.

7.3.1 Swedish 'state individualism'

The perception of Sweden as an exceptional solidary nation has been questioned by Berggren and Trägårdh (2006). Contrary to the prevailing *folkhem* slogan and official political rhetoric, the two historians see the Swedish welfare state as grounded in an outright individualistic attitude. Using the terminology of Emile Durkheim, they describe Sweden as a *Gesellschaft* whose members are driven by a marked desire for radical independence in economic as well as social terms (ibid.:10). Interestingly, it is the welfare state itself that promotes this individualism; by freeing the individual

[11] The shares in HUNT Biosciences Ltd are divided between the NTNU (34%), the Central Norwegian Health Authority (33%), and the Nord-Trøndelag County (33%).

[12] The foundation of this concept was laid in Per Albin Hansson's famous *Folkhemmet* speech in 1928.

from other people's beneficence, it secures an equal standing for all citizens.[13] Hence, what is most unusual about the Swedish society is not its social collectivism, but a peculiar alliance between the state and the individual citizen that Berggren and Trägårdh describe as 'state individualism' (ibid.:51). It builds on a specific relationship of trust between the state as provider of welfare (e.g. health services) and citizens who secure their independence by contributing to this system to some extent. By taking the traditionally close relationship of research and health care system in Sweden into account, the concept of state individualism can be applied to the issue of participation in biobank research. In particular, participation appears 'as corollary to benefiting from the healthcare system…Researchers and patients should be seen as a team, each doing their share to promote the common good of improved health' (Forsberg et al. 2009:4).

However, it should be noticed that the balance between the state, representing the interests of the community, and the individual, as beneficiary of and contributor to this system, is still a delicate one. For example, the Swedes' trust in the confidential storage and application of their samples was shaken by the access to the PKU register gained by the police, which led to the arrest of the murderer of the Swedish minister of foreign affairs, Anna Lindh (Ansell and Rasmusson 2008; Beier 2009). Although the National Board of Health and Welfare declared this unlawful, they also argued that society's interest in crime prevention might sometimes override the integrity of the individual (Socialstyrelsen 2003). However, this incident showed that trust in the actions of the state is not unlimited. Whereas in 2003 (before the police accessed the PKU register) only 17 Swedes requested removal of their samples from the PKU register, in the following year 445 citizens wanted their samples to be discarded (Hansson and Björkman 2006).

7.3.2 The Norwegian *dugnad* tradition

In Norway, the rhetorical framework of participation in biobank research is based on the concept of *dugnad*. This specifically Norwegian concept describes a voluntarily and collectively performed action. *Dugnad* dates back to nineteenth-century farming traditions when the members of the community joined together to help a fellow citizen with work that he/she could not accomplish alone. Participation in a *dugnad* did not result in any personal gain, only a meal served by the host. This established a system of reciprocity in that people participating in a *dugnad* could expect this support to be returned in case of need. The analogy with the *dugnad* tradition is deliberately expressed in the official pronouncements of the HUNT studies: 'The whole project displayed the imprint of a *dugnad* in which a whole province was the participant' (Carlsen 2009).

However, in recent years people's readiness to engage in voluntary collective work has been significantly reduced as a result of structural individualism. Thus the adoption of *dugnad* as a national word in 2004 can be seen as a 'reanimation effort' to counteract this unwelcome (from the community's perspective) tendency. Therefore the appeal to *dugnad* is not neutral in moral terms, but displays a form of 'normative recruitment', i.e. participation in human tissue research is presented as 'the right thing to do' (Ursin and Solberg 2008:109).

However, some limits on the *dugnad* analogy as a form of 'spontaneous solidarity' (ibid.:99) become apparent in its application to biobank research. Whilst researchers tend to take donor solidarity for granted, this is not yet sufficient to ensure high participation rates. In fact, the HUNT studies show a continuous decline in participants over the three projects (Holmen et al. 2004). Although premature explanations should be treated with caution, some factors that may have contributed to this decline can be identified. First, compared with the original *dugnad* activity, an individual's refusal to participate in biobank research as a health *dugnad* is obviously less

[13] See Esping-Andersen's well-known typology of welfare state regimes (Esping-Andersen 1990).

disastrous than in the context of farming; whereas a non-cooperative farmer would be left alone with his work, a patient who refuses to donate samples will still be helped in case of need. The reciprocity of the original *dugnad* system is also challenged by other factors in the field of biobanking; whilst helping in a *dugnad* was originally limited to small communities, this is not true for biobank participation. First, the mutual visibility of people contributing to a health *dugnad* is hampered by the size and character of biobank projects. Secondly, people might have difficulty in perceiving their participation as *dugnad* since they do not all gather at once for a clearly defined goal. Thirdly, if the benefit of biobanking lies in an 'increased focus on personal health, private health services and personalized medicine', there is some danger that 'in a future scenario...the social aspect of biobanks may become less important, and potentially lead to reduced support for biobanks and health surveys in the population' (NTNU 2004:18). Fourthly, since the benefits of biobanking can rarely be defined in advance, the incentives for people to support this strand of research are mainly a matter of trust (Skolbekken et al. 2005:345).[14] Finally, whilst the initial *dugnad* system was built on the equality of farmers' need for help, this is not necessarily true for the relation between donors and researchers: although the former are repeatedly asked to donate samples to researchers, this does not result the latter's being under any direct moral obligation.

7.4 **Conclusion**

The analysis of the conceptual framework of biobank participation in Sweden and Norway presented in this chapter has shown that individual and collective interests are not per se mutually exclusive in the field of biobank research. For this reason, the prevailing dichotomy between liberal individualism and the communitarian value of solidarity can be questioned. The official pronouncements of the government and research authorities in both countries take it for granted that individuals perceive their contribution to the common good as automatically serving their own interests as well. To explain this assumption, Swedish state individualism and the Norwegian *dugnad* tradition have been introduced as 'bridging concepts' between a purely communitarian framework on the one hand and a liberal approach on the other. In particular, both concepts bear on a system of reciprocity—either between the state and the individual, as in Sweden, or amongst the citizens themselves, as in Norway. However, although both countries exhibit high participation rates, analysis of the Swedish and Norwegian motivational frameworks revealed some problems which allow more general conclusions to be drawn regarding the prerequisites of participation in biobank research.

Firstly, the concept of reciprocity (see Chapter 5, section 5.6), which is at the core of the Swedish state individualism and the Norwegian *dugnad* system, cannot be taken as self-evident in modern pluralistic societies. For example, the mistrust engendered by the access to the PKU register gained by the Swedish police shows that participation is not only a question of suitable motivational predispositions of the individual, but needs to be embedded in fair and transparent regulations which the donor can trust. However, judging by participation rates, donor trust in Sweden and Norway is actually high. One reason for this can be found in the close relation between the public health-care system and medical research. The fact that the latter serves improvement of the former makes it reasonable for the individual to contribute to this system. However, if research endeavours are disengaged from health-care applications, this readiness can no longer be taken for granted. Furthermore, since empirical studies have revealed that the potential for commercialization of research is perceived as disturbing by Swedish and Norwegian citizens, this has to be

---

[14] For suggestions on how trust in biobank research can be engendered see Chapter 5, section 5.6.

taken into account as an additional feature in shaping people's willingness to participate in biobank research (see Chapter 5, section 5.4.1).

Secondly, if extensive biobank participation is not only a matter of specific motivational prerequisites of donors but also a question of trustworthy regulation and consequent review by responsible authorities, this finding might also be important for the regulation of biobank research in other countries. The traditional values of solidarity and reciprocity which facilitate the alignment of individual and collective interests are not unique to the Scandinavian countries. If the regulating authorities succeed in addressing potential donors as members of the overall community by taking the importance of personal incentives into account, this might encourage people to contribute to biobank research.[15] For this reason, the dichotomized thinking between individualism and solidarity should be replaced by efforts for a stable balance between the interests of the individual and of the community.

Finally, in this context, it is important to note that the alignment of individual and collective interest needs to be more than rhetoric but must also attain some relevance in practice. For example, the establishment of HUNT Biosciences Ltd in Norway as a 'for-profit-for-public company' makes the interplay between personal and communal benefits directly obvious to donors. That this is by no means a self-evident feature can be seen in the case of the Swedish Medical Biobank in Umeå. Its commercialization failed because of fundamental conflict between the county of Västerbotten and the university on the one hand and the researchers at the other, resulting in uncertainties and loss of confidence for both investors and donors (Laage-Hellmann 2003).

References

Act on Medical and Health Research (2008). Available at: http://www.helsebiblioteket.no/Nyhetsarkiv/32529.cms (accessed 25 September 2009).

Adami, H.O. ***et al.*** (2002). Tretton forskare protesterar mot nya biobankslagen (Thirteen researchers protest against the new Biobank Act). *SVEPET, Medlemsblad för Svensk Epidemeologisk Förening*, **1**, 8–10.

Ansell, R. and Rasmusson, B. (2008). A Swedish perspective. *Biosocieties*, 3, 88–92.

Beier, K. (2009). Between individualism and solidarity. Biobanking in Sweden. In K. Dierickx and P. Borry (eds), *New challenges for biobanks. Ethics, law and governance*, pp.49–64, Antwerp: Intersentia.

Berggren, H. and Trägårdh, L. (2006). *Är svensken människa? Gemenskap och oberoende i det moderna Sverige* (Is the Swede human? Community and autonomy in modern Sweden). Stockholm: Norstedts.

Biobanks in Medical Care Act (2002). Available at: http://www.notisum.se/rnp/sls/lag/20020297.htm (accessed 13 May 2009).

Carlsen, T. (2009). *HUNT—25 år for folkehelsen.* (HUNT—25 years for public health). Available at: http://www.universitetsavisa.no/ua_lesmer.php?kategori=nyheter&dokid=4a76cd648dbd71.32093150 (accessed 18 October 2009).

Carlsson, I. (2008). *Potential of improving quality of life (Interview on LifeGene website).* Available at: http://lifegene.ki.se/about/interviews/carlsson_en.html (accessed 18 October 2009).

Data Protection Directive (1995). *Directive 95/46/EC of the European Parliament and of the Council of 24 October 1995 on the protection of individuals with regard to the processing of personal data and on the free movement of such data.* Available at: http://ec.europa.eu/justice_home/fsj/privacy/docs/95-46-ce/dir1995-46_part1_en.pdf (accessed 23 March 2009).

Esping-Andersen, G. (1990). *The three worlds of welfare capitalism.* Cambridge: Polity Press.

Forsberg, J.S., Hansson, M.G., and Eriksson, S. (2009). Changing perspectives in biobank research: from

[15] Admittedly, this is one of the most challenging tasks in the governance of biobank research. For example, the French legal framework takes insufficient account of donor participation (see Chapter 11, section 11.2).

individual rights to concerns about public health regarding the return of results. *European Journal of Human Genetics*, **17**, 1544–9.

FUGE nyhetsbrev (2007). Folk ønsker å bidra (People wish to contribute). *FUGE nyhetsbrev*, 2, 6–7. Available at: http://www.forskningsradet.no/servlet/Satellite?c=Page&cid=1226994105732&pagename=fuge%2FHovedsidemal (accessed 13 October 2009).

Gentekniknämnden (2004). *Konferensrapport: DNA-Analyser, DNA-register och biobanker.* (Conference report: DNA analysis, DNA register and biobanks). Available at: http://www.v-a.se (accessed 15 May 2009).

Hansson, S.O. and Björkman, B. (2006). Bioethics in Sweden. *Cambridge Quarterly of Health Care Ethics*, **15**, 285–93.

Helgason, H.H. (2007). Consent and population genetic databases: a comparative analysis of law in Iceland, Sweden, Estonia and the UK. In M. Häyry *et al.* (eds) *The ethics and governance of human genetic databases. European perspectives*, pp. 97–107, Cambridge: Cambridge University Press.

Hoeyer, K.L. (2004a). *Biobanks and informed consent. An anthropological contribution to medical ethics.* Available at: http://www.diva-portal.org/umu/abstract.xsql?dbid=358 (accessed 18 February 2009).

Hoeyer, K.L. (2004b). Ambiguous gifts: public anxiety, informed consent and biobanks. In R. Tutton and O. Corrigan (eds) *Genetic databases. Socio-ethical issues in the collection and use of DNA*, pp. 97–116, London: Routledge.

Holmen, J., Midthjell, K., Krüger, Ø., *et al.* (2003). The Nord-Trøndelag Health Study 1995-97 (HUNT 2): objectives, contents, methods and participation. *Norsk Epidemiologi*, **13**, 19–32.

Holmen, J., Kjelsaas, M.B., Krüger, Ø., *et al.* (2004). Befolkningens holdninger til genetisk epidemiologi illustrert ved spørsmål om fornyet samtykke til 61.426 personer–Helseundersøkelsen i Nord-Trøndelag (HUNT) (The population's attitudes towards genetic epidemiology illustrated by the issue of renewed consent for the 61,426 person health study (HUNT)). *Norsk Epidemiologi*, **14**, 27–31.

Jansen, B. (2004). *Rechtliche und ethische Aspekte von DNA-Banken im internationalen Vergleich (Legal and ethical aspects of DNA banks as compared to international standards).* Munich: GRIN Verlag.

Johnsson, L., Hansson, M.G., Eriksson, S., *et al.* (2008). Patient's refusal to consent to storage and use of samples in Swedish biobanks: cross sectional study. *British Medical Journal*, **337**, a345.

Knoppers, B.M. and Ruth, C. (2005). Human genetic research: emerging trends in ethics. *Nature Reviews Genetics*, **6**, 75–9.

Kvalvaag, H. (2004). *Biobankloven hindrer kreftforskningen.* (The Biobanks Act hampers cancer research). Available at: http://www.forskning.no/artikler/2004/oktober/1098103860.17 (accessed 11 October 2009).

Laage-Hellmann, J. (2003). Clinical genomics companies and biobanks. The use of biosamples in commercial research on the genetics of common diseases. in M.G. Hansson and M. Levin (eds) *Biobanks as resources for health*, pp. 51–90, Uppsala: Universitetstryckeriet Uppsala.

Landstingen (2004). *Biobanken–resurs för livet* (Biobank–a resource for life). http://www.biobanksverige.se/getDocument.aspx?id=150 (accessed 9 March 2009).

Landstingen (2005). *Landstingens gemensamma biobanksdokumentation. Informations- och samtyckesordning* (The county councils' common documentation on biobanks: rules for information and consent). Available at: http://www.biobanksverige.se/getDocument.aspx?id=37 (accessed 22 April 2009).

LifeGene Ethics Group (2009). *LifeGene Ethics Policy*. Available at: http://lifegene.ki.se/ethical_issues/documents/090331LifeGeneEthicsPolicyv32.pdf (accessed 13 October 2009).

LifeGene Informationsbroschyr (2009). *Vad är LifeGene och varför ska jag delta?* (What is LifeGene and why should I participate?). Available at: https://www.lifegene.se/PageFiles/277/Info%20dokument/lifegene_broschyr_16.pdf (accessed 3 November 2009).

LifeGene Newsletter (2009). Available at: http://lifegene.ki.se/activities/documents/LifeGenenewsletterjun-09.pdf (accessed 16 November 2009).

Nilstun, T. and Hermerén, G. (2006). Human tissue samples and ethics. Attitudes of the general public in Sweden to biobank research. *Medicine, Health Care and Philosophy*, **9**, 81–6.

Nilsson, A. and Rose, J. (1999). Sweden takes steps to protect tissue banks. *Science*, **286**, 894.

Nobel, S. (2008). *Biobanks–integration of human information to improve health. Report of the Committee for Research Infrastructures and the Scientific Council for Medicine at the Swedish Research Council.* Available at: http://www.vr.se/download/18.61c03dad1180e26cb8780004896/Spr%C3%A5kgranskad+Biobank_report_SRC_v2.pdf (accessed 30 September 2009).

NOU (2005). *God forskning—bedre helse.* (Good research—better health). Available at: http://www.regjeringen.no/Rpub/NOU/20052005/001/PDFS/NOU200520050001000DDDPDFS.pdf (accessed 13 October 2009).

NTNU (2004). *Foresight analysis medical technology: health surveys and biobanking.* Available at: http://www.bioethics.ntnu.no/docs/Foresightrapport.pdf (accessed 9 November 2009).

Personal Data Act (1998). Available at: http://www.notisum.se/rnp/sls/LAG/19980204.htm (accessed 11 February 2009).

Rynning, E. (2003). Public law aspects on the use of biobank samples: privacy versus the interest of research. In M.G. Hansson and M. Levin (eds) *Biobanks as resources for health*, pp.91–128, Uppsala: Universitetstryckeriet Uppsala.

Skolbekken, J.-A., Ursin, L.Ø., Solberg, B., Christensen, E., and Ytterhus, B. (2005). Not worth the paper it's written on? Informed consent and biobank research in a Norwegian context. *Critical Public Health*, **15**, 335–47.

Socialdepartementet (2008). *Kommittédirektiv: Översyn av lagen (2002:297) om biobanker i hälso- och sjukvården m.m.* (Review on the Biobanks in Medical Care Act). Available at: http://www.riksdagen.se/webbnav/index.aspx?nid=10&dok_id=DIR2008:71 (accessed 7 April 2009).

Socialstyrelsen (2002). *Socialstyrelsens föreskrifter och allmänna råd om biobanker i hälso- och sjukvården m.m.* (The National Board of Health and Welfare's direction and general recommendation on biobanks in health care and medical ward). Available at: http://www.sos.se/sosfs/2002_11/2002_11.pdf (accessed 18 March 2009).

Socialstyrelsen (2003). *Tillhandahållande av vävnadsprover vid utredningen av brott.* (Delivery of tissue samples for criminal investigation). Available at: www.sos./se/fulltext/107/2003-107-16/2003-107-16.htm (accessed 7 January 2009).

Solum Steinsbekk, K. et al. (2009). From idealism to realism. Commercial ventures in publicly funded biobanks. In K. Dierickx and P. Borry (ed.). *New challenges for biobanks. Ethics, law and governance*, pp.137–51, Antwerp: Intersentia.

Steinmann, M. (2009). Under the pretence of autonomy: contradictions in the guidelines for human tissue donation. *Medicine, Health Care and Philosophy*, **12**, 281–9.

Trägårdh, M. and Ringman M. (1999). Ditt liv är till salu (Your life is for sale). Available at: http://www.aftonbladet.se/nyheter/9904/10/cell.html (accessed 21 January 2009).

Ursin, L.Ø. et al. (2008). The informed consenters: governing biobanks in Scandinavia. In H. Gottweis and A. Petersen (eds). *Biobanks. Governance in comparative perspective*, pp.177–93, London: Routledge.

Ursin, L.Ø. and Solberg, B. (2008). When is normative recruitment legitimate? *Etikk I praksis. Nordic Journal of Applied Ethics*, **2**, 93–113.

Wendler, D. (2002). What research with stored samples teaches us about research with human subjects. *Bioethics*, **16**, 33–54.

Wilson, S. (2004). Population biobanks and social justice: commercial or communitarian models? *Trames—Journal of the Humanities and Social Sciences*, **8**, 80–9.

Part II

The legal regulation of human tissue research

Chapter 8

Law, ethics, and human tissue research: integration or competition?

José Miola

8.1 Introduction

The Human Tissue Act 2004 received Royal Assent in England and Wales in November 2004 and has replaced the previous statutes governing the retention, storage, and use of human organs and tissue (including for research purposes), which have therefore been repealed. As the Act's own explanatory notes make clear, the legislation was a consequence of several medical scandals involving human organs and tissue, in particular where it was 'established that organs and tissue from children who had died had often been removed, stored and used without proper consent' (Human Tissue Act Explanatory Notes 2004:para 5). Moreover, this practice was found to have been widespread in the past (ibid.). The various reports and inquiries into these scandals recommended that the law was 'not comprehensive, nor as clear and consistent as it might be for professionals or for the families involved', and that the law should be changed (ibid.:para 4; see also Chapter 3, section 3.4, of this volume).

Therefore the new legislation was aimed at providing a cogent and consistent legislative framework that would tighten the law and rectify the perceived failings of the old system of governance. Perhaps unsurprisingly, the concept of consent was identified as the 'fundamental principle' that would underpin the Act (Human Tissue Act Explanatory Notes 2004:para 4). However, alongside this underlying philosophy, the Act also developed a systemic change, creating a Human Tissue Authority whose role was 'intended to rationalize existing regulation of activities like transplantation and anatomical examination, and will introduce regulation of other activities like post-mortem examinations, and the storage of human material for education, training and research' (ibid.).

In this chapter I explore the background to the Act, as well as the Act itself. I shall demonstrate that the Act is an understandable and predictable response to the factors that brought about its enactment, on both substantive and systemic levels. However, I shall also argue that the Act not only reflects dissatisfaction with the previous legal rules, but also represents a failure on the part of medical ethics, and in particular the way in which ethics are presumed to regulate medical professionals and researchers. In order to do this, we must first briefly examine the way in which medical ethics operates in England and Wales.

8.2 Ethical regulation in England

The regulation of the medical profession and medical professionals in England is, outwardly at least, rigid and clear. The Medical Act 1983 tasks the General Medical Council (GMC) with the responsibility for the creation and maintenance of the medical register. This means that it alone has the power to decide who may (and who may not) practice medicine in England. Section 35 of the Act provides the GMC with the further power to deliver, 'in such manner as the council think fit', advice on standards of professional conduct, professional performance, and (crucially for the purposes of this chapter) medical ethics. In the GMC's own guidance, it is made clear that its advice is intended to be authoritative. Therfore doctors are informed that if a portion of GMC guidance contains the words 'you must', then it is compulsory to adhere to the guidance. Furthermore, the guidance represents the standard by which the doctor will be judged, and that '[s]erious or persistent failure to follow this guidance will put your registration [as a medical practitioner] at risk' (GMC 2006).

Therefore it might be thought that, in an area so rich with ethical controversy, the ethical guidance from the GMC relating to the use of human organs and tissue in research would be comprehensive and up to date. Unfortunately, this is not the case at all, and has not been for some time. The last time that the GMC published a dedicated document on the subject of human tissue and organs was 1992 (GMC 1992). It warned doctors not to participate in the trade in organs from live donors, and was withdrawn in 2006. The only current mentions are to be found in generalized guidance on research, which specifies that appropriate consent must be sought before organs may be taken from donors (GMC 2002). Perhaps unsurprisingly, and in common with other areas of medical law and ethics where the GMC's advice is less than comprehensive, it cannot be said to have the monopoly on ethical guidance by professional groups that its unique status might suggest that it should (Miola 2007). Rather, the British Medical Association (BMA), which is essentially the doctors' trade union, and the Royal Colleges each publish their own guidance on ethical issues. In many cases, and in many respects, the guidance from these bodies, particularly the BMA, can be said to be superior to that of the GMC in the sense that it is more comprehensive and interactive. For example, the BMA's general guidance runs to over 800 *pages*, compared with merely 79 *paragraphs* in the GMC version (BMA 2004; GMC 2006).

If it is the role of the GMC to provide comprehensive ethical guidance to medical professionals, then in the context of human tissue and research it can only be said to have failed. However, it is noticeable that the Medical Act only provides the GMC with a *power* to advise on medical ethics, rather than a *duty*. Nevertheless, as we shall see later, the existence of disparate sources of guidance has been part of the problem that led to the enactment of the legislation. Yet, when this is combined with the fact that the medical profession in England has suffered from several scandals in the last two decades, the regulatory framework created by the Human Tissue Act 2004 becomes much more logical.

8.3 Background to the Human Tissue Act 2004

The problems created by the lack of ethical leadership shown by the GMC were exacerbated by the fact that the law before the 2004 legislation also did little to rein in medical practitioners. Indeed, the Human Tissue Act 1961, the predecessor of the Human Tissue Act 2004, can best be said to be a muddle, for two main reasons. First, section 1(1) stated that if a person, 'during his last illness', had expressed a request for parts of his body to be used for transplantation or research, the 'person lawfully in possession of the body' could authorize the removal of the organs post-mortem. However, and amazingly, it failed to provide a definition regarding who would be in 'lawful possession' of the body (Skegg 1984)! It was generally felt that if a person died in hospital then the hospital would be in lawful possession (Lanham 1971). Secondly, and perhaps even more perplexingly, although the 1961 Act specified conditions for the removal of organs and tissue, it never provided any sanction for non-compliance. Therefore there was confusion regarding even fundamental questions such as whether any breach of the statute should be dealt with in the criminal or civil courts (Kennedy 1989).

The 'final straw' that provided the catalyst for the change in the law was a slew of medical scandals that have occurred in the UK over the past two decades, some of which involved the retention and use of organs and tissue without consent, but all of which helped to produce a lessening of trust and confidence in the medical profession. Perhaps the most famous of the medical scandals of recent decades has been that involving Harold Shipman, a general practitioner and serial killer who murdered; it was estimated by the inquiry into his conduct, hundreds of his patients (Shipman Inquiry 2005). For obvious reasons, this had the effect of reducing the public's trust in the medical profession (Miola 2007). However, there were also scandals relating to organs, in particular some taken from children and babies. During the inquiry into the paediatric deaths at Bristol Royal Infirmary (due to inadequately performed cardiac surgery), it was found that in many cases organs were removed from the patients without the consent, or even knowledge, of their parents, and in many cases were never returned to the bodies before burial (Bristol Inquiry 2001a). Yet the situation was to worsen. During the Bristol Inquiry, it also came to light that such practices were considered normal by the medical profession, and that Alder Hey Children's Hospital in Liverpool had done exactly the same. Another inquiry was set up, chaired by the prominent lawyer Michael Redfearn, and again the medical profession was seen to have removed and retained organs from children without the knowledge or consent of the parents (Hall and Lilleyman 2001; Royal Liverpool Children's Inquiry 2001). When this was found to be common practice nationally, a Retained Organs Commission was established to oversee the return of retained tissue, and also to propose new legislation and guidance for the future. From this emanated the consultations that brought about the Human Tissue Bill, which became the Human Tissue Act 2004.

What can be seen is that the journey to the Human Tissue Act 2004 began, and was driven by, an *ethical and legal* failure to regulate the medical profession, which in turn acted in a way that society found deeply offensive. The only conclusion that can be drawn is that laissez-faire ethics and law had allowed medical practitioners to act in the way that they did. This background explains, almost completely, why the 2004 Act is as it is. In the next section I shall continue by describing the Act itself.

8.4 **The provisions of the Human Tissue Act 2004**

The Human Tissue Act 2004 emphasizes the importance of two key themes that constitute obvious responses to the problems and medical behaviour described in the last section: autonomy (of donors) and accountability (for medical professionals). That these were to be the two bedrocks of the new Act was made clear by Alan Milburn, then Secretary of State for Health, in the House of Commons:

> ...*the law will be changed to enshrine the concept of informed consent.* The existing law in this area has become outdated. *The Human Tissue Act 1961 does not even contain penalties for breaches of its provisions.* The law has ill served bereaved parents in our country and causes confusion for staff. It must be changed (Hansard 2001).

I begin by considering the first of these. The Act states that it shall be unlawful to remove, use, or store organs unless this is done by a licensed clinic for a purpose authorized by the Act (Human Tissue Act 2004:section 1). One of the listed purposes is '[r]esearch in connection with disorders, or the functioning, of the human body' (ibid.:Schedule 1, Part 1). However, this is on the condition that this is done with the 'appropriate consent' of the donor. As Mr Milburn intimates, 'appropriate' consent means, for competent donors, informed consent. This much is uncontroversial, although the law relating to what actually constitutes informed consent in England is rather unclear (Miola 2009). However, the desire to prioritize autonomy is somewhat compromised by the need to ensure an adequate supply of organs and tissue for transplantation and research purposes. This creates a particular tension when it is applied to the treatment of tissue and organs provided by dead donors, and this too is evident in the Act. Thus, the Act requires a three-part procedure which allows for the removal, storage, and use of organs and tissue from the dead.

The first is where the person has previously signalled his/her consent for this to occur, such as by creating a living will (Human Tissue Act 2004:section 3(6)(a)). If this has not occurred, appropriate consent may be obtained from a nominated representative of the deceased, if one was appointed (ibid.:section 3(6)(b)). Strangely, a person may appoint more than one nominated representative. If this is the case, then they are held to act either jointly or severally, unless it is made clear by the appointer that they are to act jointly (ibid.: section 4(6)). The final method for gaining consent, which operates if the other two do not apply, is that it may be provided by someone with a 'qualifying relationship to the deceased' (ibid.:section 3(6)(c)). The list of relationships is as follows:

(a) spouse or partner;

(b) parent or child;

(c) brother or sister;

(d) grandparent or grandchild;

(e) child of a person falling within paragraph (c);

(f) stepfather or stepmother;

(g) half-brother or half-sister;

(h) friend of long standing (ibid.: section 27(4)).

The person asking for the donation must work his/her way down the list, seeking the consent of the highest-ranked person on the list who may reasonably be found. If more than one person meets the criteria, they should be given equal ranking (ibid.:section 27(5)), although it is sufficient to obtain the consent of only one of them (ibid.:section 27(7)). Needless to say, this is the most controversial of the methods of obtaining consent, for two reasons. The first is that it does not require the direct consent of the donor or a person whom he/she has specifically appointed to make decisions. In a statute whose basis is the principle of autonomy, and in which a full third of the content is made up of procedures and rules relating to consent, this may be seen as inconsistent with the overall philosophy. Secondly, the list of qualifying relationships may only be described as ad hoc, and no provision is made for any recognition that the deceased may have been closer to a lower-ranked individual. Therefore there is no mechanism for changing the ranking if the deceased was estranged from his/her parents but close to his/her grandparents. As argued earlier, the three-part procedure for obtaining consent can only be seen as a compromise between a desire for autonomy and a need for organs and tissue.

In terms of accountability (Mr Milburn's second bedrock), the Act had a simple problem to solve, and does so in the most obvious fashion. The 'problem', as mentioned in the previous section, was that the previous legislation actually specified no sanction for non-compliance. The resolution is that the new statute should provide some sanction. To this end, the Human Tissue Act 2004 states that anyone removing, storing, or using organs or tissue without authorization may face criminal charges, resulting in a fine and/or a term of imprisonment not exceeding three years (ibid.:section 5(7)).

What is evident in the Act, then, is that the twin concerns relating to autonomy and accountability—the drivers behind its creation in the first place—have been addressed in the statute.

How effective or philosophically consistent it has been done is open to conjecture, but attempts have been made to address the lacunae in the previous legislation. To strengthen its ability to regulate, the Act also establishes the Human Tissue Authority, which is considered in the next section.

8.5 The Human Tissue Authority

In order to support and deliver the regulatory framework envisioned and required by the legislation, the Act established the Human Tissue Authority as an overseeing body. The remit of this authority, outlined in section 14 of the Act, is wide and includes virtually all issues relating to the removal, storage, use, and even disposal of human organs and tissue. If the remit of the Authority is wide, then the same can be said for the amount of discretion it has at its disposal. This is evident in the way that the Act defines its general functions.

> The Authority shall have the following general functions
> (a) maintaining a statement of the general principles which it considers should be followed
> (i) in the carrying-on of activities within its remit, and
> (ii) in the carrying-out of its functions in relation to such activities;
> (b) providing in relation to activities within its remit such general oversight and guidance as it considers appropriate;
> (c) superintending, in relation to activities within its remit, compliance with
> (i) requirements imposed by or under Part 1 or this Part, and
> (ii) codes of practice under this Act;
> (d) providing to the public, and to persons carrying on activities within its remit, such information and advice as it considers appropriate about the nature and purpose of such activities;
> (e) monitoring developments relating to activities within its remit and advising the Secretary of State, the National Assembly for Wales and the relevant Northern Ireland department on issues relating to such developments;
> (f) advising the Secretary of State, the National Assembly for Wales or the relevant Northern Ireland department on such other issues relating to activities within its remit as he, the Assembly or the department may require (ibid.: section 15).

As can be seen from the above, the functions of the authority are various and critical to the operation of the Act's governance framework. Paragraph (a) tasks the Authority with maintaining the general principles that must be followed. This is done through the creation and updating of a Code of Practice relating to the Act. Paragraphs (b) and (c) emphasize the duty of the Authority to oversee and guide those carrying out activities under the auspices of the Act, while paragraphs (d), (e), and (f) outline the Authority's duties to the public and to government ministers (including a duty to advise). Furthermore, as section 16 makes clear, in order to undertake any activity under the Act, it is first necessary to obtain a licence from the Authority, which also retains the right to revoke it if the provisions are not adhered to.

Therefore it is clear that the Authority is envisioned by the Act as a powerful and influential organization, and this is not without parallel in England. Indeed, the concept of an overarching 'Authority' supervising a legislative framework also exists with regard to new reproductive technologies and embryo research with the Human Fertilization and Embryology Authority, which performs what is essentially the same role. The principle advantage, which translates across to the Human Tissue Authority, is that it provides a living breathing organism that, unlike a statute, can quickly respond to changes in circumstance and technological advancement, thus keeping up to date and maintaining relevance. Indeed, fixed statutes in dynamic areas of law can rapidly prove cumbersome and clumsy. For that reason, if no other, the Human Tissue Authority should not be seen solely as a demonstration of Parliament's lack of faith in medical professionals and scientists. Nevertheless, and in particular given the reasons for the Human Tissue Act itself, something of a lack of faith is understandable given the background to the Act, and the removal of self-regulation and the imposition of an overarching Authority cannot be said to be mere coincidence. In this sense, the establishment of the Authority fits perfectly the several requirements of the purpose of the Act.

8.6 **Integration, not competition?**

Therefor it might be said that the Human Tissue Authority is not just the living breathing embodiment of the law but also, in effect, the definitive statement of ethical standards. As I have previously mentioned in this chapter, the ethical regulation of medical and biomedical practice in England comprises a fragmented network of different bodies, each purporting to advise its members (who may be members of several of these organizations at once). While the GMC is, in theory, at the head of this group, all too often it does not provide authoritative and comprehensive guidance that renders that from other bodies superfluous. Indeed, the first of the medical 'scandals' that provided the impetus for the creation of the Act recognized the fact that the regulation of medical professionals lacked cohesion. The inquiry into the events at Bristol Royal Infirmary was chaired by Professor Sir Ian Kennedy, a leading academic medical lawyer. In the report, it was noted that the events that occurred at Bristol were the result of a process, and that a part of that process was the creation of separate 'silos' of responsibility followed by a fragmentation of guidance leading to a lack of coordination of standards. Thus guidelines were found to 'appear from a variety of bodies giving rise to confusion and uncertainty' (Bristol Inquiry 2001b:17).

Rather than provide for 'better' regulation, this preponderance of guidance essentially cancelled itself out, paradoxically leading to *less* effectiveness as each group assumed that another was responsible for setting and maintaining standards. Again, this was found to be evident amongst Bristol's paediatric heart surgeons.

> The SRSAG [Supra Regional Services Advisory Group] thought that the health authorities or the Royal College of Surgeons were doing it; the Royal College of Surgeons thought that the SRSAG or the trust were doing it, and so it went on. No one was doing it. We cannot say that the external system for assuring and monitoring the quality of care was inadequate. There was, in truth, no such system (ibid.:192).

The result, according to the inquiry, was that therefore 'no such standards exist' and that 'healthcare professionals may not know which standards should be followed, or what status or authority the standards have' (ibid.:383). The story of Bristol, then, was not just one of a failure of law. It was also one of a failure of medical ethics (Miola 2007). One of the main recommendations of the report was that there should be coordination of standard setting and maintenance.

> Currently, there are a large number of bodies involved in the activities which together constitute regulation. They include the new Nursing and Midwifery Council, the GMC, the proposed new body which will regulate the professions allied to medicine, the Royal Colleges, the various professional associations, the Department of Health, health authorities and trusts. Each operates in its own sphere, with, historically, little collaboration or co-operation. The various activities must be brought together and properly co-ordinated. The role of the various bodies must be clearly identified. And all of the bodies should be brought under the overall leadership of one overarching body (Bristol Inquiry, 2001b:315).

The Human Tissue Act, through the Human Tissue Authority, establishes precisely the sort of regime that is envisaged by Kennedy in the inquiry's report. Any activity governed by the Act (removal, storage, and use of organs and tissue for a scheduled purpose) may only be undertaken following the granting of a licence by the Human Tissue Authority. If there are irregularities, or the Act is not complied with, the Authority may remove that licence, thus preventing further work from taking place. Therefore what can be seen is that there is a single body upon which lies the responsibility for the setting and maintenance of standards. Moreover, the fact that this body has the power to advise government and keep the guidance up to date means that it is an overarching body that performs both a legal and an ethical function. In other words, the legal and ethical standards are essentially homogenized, and the lines of distinction between them are removed. Therefore the Human Tissue Authority can be said to *defragment* the ethics in this area.

8.7 **Conclusion**

As I mentioned at the beginning of this chapter, the Human Tissue Act 2004 was created and passed in response to some specific problems. Essentially, the medical profession had lost the trust of the public and legislators as a result of a series of scandals, many concerning the removal, storage, and use of human tissue and organs. Moreover, the lack of accountability and sanctions inherent in the law relating to human tissue and organs at that time were put into stark focus by the scandals, and it was clear that new legislation was required. Given this, the substantive provisions of the new Act are both predictable and understandable. As I have shown, the Act is an active response to the perceived deficiencies in the old law in that it both emphasizes and centralizes the role of consent (thus in theory prioritizing autonomy, although it might take more than this to actually do so (see Chapter 2)), and it also introduces criminal sanctions for non-compliance with the terms of the statute (thus creating a system of accountability).

However, what is perhaps of even more interest is the new regulatory structure that the Act provides. The establishment of the Human Tissue Authority allows the statute to live and breathe and, effectively, to update itself. As I have sought to demonstrate, its powers are wide-ranging, covering the licensing of establishments seeking to operate under the Act, the right to compose and update guidance on good practice, and the right to advise the government. It would not be too much of a stretch, I believe, to refer to this body as the Human Tissue Act's judiciary. However, what is also noteworthy is the way in which the Authority is envisioned as being the central hub of guidance relating to work performed under the Act. This is because, essentially, it also has the effect of replacing medical ethics. The Act's substantive legal rules are underwritten by a system of ethical guidelines which, nevertheless, must be followed if a license is to be retained. The Authority and its guidance thus defragments ethical discourse—once again, as I have demonstrated, perhaps in response to specific criticisms of the old model (in this case the Bristol Royal Infirmary Inquiry Report).

Whether or not these are the reasons for the regulatory structure is somewhat beside the point. Indeed, we cannot be sure that it is due to the fact that the establishment of an 'Authority' to oversee the running of an Act is not unprecedented in English medical law (as the Human Fertilization and Embryology Act makes clear with the Human Fertilization and Embryology Authority). Nevertheless, what can be said is that if it is a coincidence, it is a felicitous one as it solves a major problem with the old regime covering human tissue and organs. The new law is not perfect, but it has at least sought to solve the issues with the previous law that attracted the majority of the criticism. Furthermore, by defragmenting medical ethics, it also manages to make medical ethics a *part of* the law, rather than, if anything, a competitor. If nothing else, this must be seen as a positive development.

References

BMA (2004). *Medical ethics today: The BMA's handbook of ethics and law.* London: BMA Books.

Bristol Inquiry (2001a). *Bristol Inquiry Interim Report.* Available at: http://www.bristol-inquiry.org.uk/interim_report/annexb1.htm (accessed 11 September 2009).

Bristol Inquiry (2001b). *Learning from Bristol: the report of the public inquiry into children's heart surgery at the Bristol Royal Infirmary 1984–1995.* London: The Stationery Office.

GMC (1992). *Transplantation of organs from live donors.* Available at: http://www.gmc-uk.org/guidance/archive/transplantation_live_donors_1992.pdf (accessed 10 September 2009).

GMC (2002). *Research: the role and responsibilities of doctors.* Available at: http://www.gmc-uk.org/guidance/current/library/research.asp#consent (accessed 10 September 2009).

GMC (2006). *Good medical practice.* Available at: http://www.gmc-uk.org/guidance/good_medical_practice/how_gmp_applies_to_you.asp (accessed 10 September 2009).

Hall, D. and Lilleyman, J.S. (2001). Reflecting on Redfearn: what can we learn from the Alder Hey story? *Archives of Disease in Childhood*, **84**, 455–7.

Hansard (2001). 30 January 2001. Available at: http://www.publications.parliament.uk/pa/cm200001/cmhansrd/vo010130/debtext/10130-06.htm (accessed 2 November 2009).

Human Tissue Act 2004 (2004). Available at: http://www.opsi.gov.uk/ACTS/acts2004/ukpga_20040030_en_1 (accessed 13 October 2009).

Human Tissue Act 2004 Explanatory Notes (2004). Available at: http://www.opsi.gov.uk/acts/acts2004/en/ukpgaen_20040030_en_1 (accessed 13 October 2009).

Kennedy, I. (1989). Further thoughts on liability and the Human Tissue Act 1961. In Kennedy, I. *Treat me right: essays on medical law and ethics*, pp. 225–37. Oxford: Oxford University Press.

Lanham, D. (1971). Transplants and the Human Tissue Act 1961. *Medicine, Science and the Law*, **11**, 16–24

Miola, J. (2007). *Medical ethics and medical law: a symbiotic relationship.* Oxford: Hart.

Miola, J. (2009). On the materiality of risk: paper tigers and panaceas. *Medical Law Review*, **17**, 76–108

Royal Liverpool Children's Inquiry (2001). *Royal Liverpool Children's Inquiry Report (The Redfearn Report).* London: The Stationery Office.

Shipman Inquiry (2005). Available at: http://www.the-shipman-inquiry.org.uk/home.asp (accessed 11 September 2009).

Skegg, P.D.G. (1984). *Law, ethics, and medicine.* Oxford: Oxford University Press.

Chapter 9

Legal paradigms of human tissues

Remigius N. Nwabueze

9.1 Introduction

Taking a paradigm as a conceptual model of analysis, it is possible to adumbrate three overlapping ways in which legal analyses of, and reflections on, human tissues are often projected.[1] For want of better terms, these are characterized as the traditional, critical, and modern approaches, although these modes of scholarship on human tissues do not occupy watertight compartments. However, significant differences exist among the tripodic analytical frameworks for human tissues. This justifies their compartmentalization in the way suggested in this chapter. However, since each of the three analytical paradigms on human tissues could be the subject of a separate chapter, I propose to give only their outlines. Consequently, the analysis presented merely attempts to explore the essence of a particular approach and its relation to other approaches. It should be noted that this chapter has not attempted to integrate the impact of legislation on human tissues into the conceptual frameworks highlighted above. Human tissues are covered by a wide range of legislation, including that on anatomy and dissection, organ donation and transplantation, fertilization and reproductive technology, tissue banks, disposal of hospital wastes containing human tissues, and criminal law. For two reasons, none of the three paradigms identified above focuses on these statutes. First, some statutory aspects of human tissues have been examined elsewhere (Nwabueze 2002). Secondly, adequate analysis of the diversity and multiplicity of legislation on human tissues (both nationally and internationally) cannot, arguably, be captured in a chapter-length work. Where necessary, however, references have been made to relevant legislation.

9.2 Traditional approach

Human tissue scholarship is traditionally characterized by common law's treatment of cadavers as 'things' outside the zone of property protection. This is generally known as the 'no-property rule'. While whole and buried cadavers provided the context for the no-property rule, it has nevertheless been applied to excised human tissues.[2] Historical justifications for the no-property rule are based on the biomedical and commercial inutility of cadavers and human tissues. In the medieval period, for instance, burial or cremation was the unquestionable destination of a corpse.

[1] 'Human tissue' is used in a broad sense to include all excised parts of a human body, including the cells and products of a human body (for similar usage see Nuffield Council on Bioethics 1995:17–19; Waldby and Robert 2006:6).

[2] The observation of Pring in *Doodeward* v. *Spence* provides a judicial example of this extension: 'There can be no property in a human body dead or alive. I go further and say that if a limb or any portion of a body is removed that no person has a right of property in that portion of the body so removed' (*Doodeward* v. *Spence* [1908] 6 CLR 406).

After death, a cadaver had no further use or value except to be interred to the ground.[3] Not even the horrible black market in dead bodies conducted by resurrectionists in the eighteenth and early nineteenth centuries could be said to have attributed a commercial value to cadavers.[4] As Scott (1981) noted, while cadavers in the hands of resurrectionists attracted considerable price from various schools of anatomy (see also the example of the 'Irish Giant' cited in Chapter 1), the transaction 'was really a black market operating outside the law or on the fringes of the law. The demand was highly specialized and limited in quantity' (ibid.:12). Thus body-snatching, despite the filthy lucre that was involved, conferred no real value on cadavers. It is only through recent biomedical applications, including organ transplantation, medical research, and experimentation, that the human body and body parts have gained an intrinsic value and utility. As rightly observed, the 'true reason for the unique worth which the body now has is the newly discovered capacity of its tissues and organs to cure disease and repair defective bodies. Its value, then, is a function of its therapeutic utility' (ibid.:14) Therefore, since a dead human body is scientifically and economically valueless, common law unsurprisingly took the view that no property interests were inherent in it.

While the no-property rule is rooted in antiquity and enjoys the advantage of clarity in its formulation[5], its legal foundations are shrouded in obscurity. The Law Reform Commission of Canada traced the rule to Sir Edward Coke's famous commentary that burials were within ecclesiastical cognisance and that the burial of cadaver was *nullius in bonis* (the property of no one) (Law Reform Commission of Canada 1992). In terms of precedent, however, the no-property rule is based on the seventeenth century *Haynes's case* [1614]. As a provenance, however, neither Coke nor *Haynes's case* provides a reliable path to the no-property rule (Matthews 1983:197–8). First, Coke's proposition was not based on any prior judicial authority; nor was it justifiable since there were things protected by common law (such as church fabric and monuments) while, at the same time, under ecclesiastical cognisance. Furthermore, *Haynes's case* is often falsely associated with the proposition that a corpse is not capable of being property. Rather, the case posited that a corpse was not capable of owning property (ibid.). Whatever the merits of its provenance, the no-property rule has survived for more than three hundred years (amongst others, it is one of the founding principles of the French bioethics approach that also regulates human tissue research; see Chapter 11).[6] It has become an established and authoritative analytical framework for cadavers in most Western legal systems (Nwabueze 2007a). These systems also recognize some limited exceptions to the no-property rule based on work and skill on cadavers and human tissues.[7] Interestingly, some civil law systems have also embraced the no-property rule.[8]

[3] However, there are indications that even in the medieval period the human body was used for anthropophagic medicine (Gordon-Grube 1988:405–9). I am grateful to Nils Hoppe for bringing this to my attention.

[4] Resurrectionists were 'body-snatchers' who specialized in stealing freshly buried dead bodies for sale to medical schools. For a good history of their activities see Richardson (2006).

[5] Blackstone lucidly commented that 'though the heir has a property interest in the monuments and escutcheons of his ancestors, he has none in their bodies or ashes; nor can he bring any civil action...against their bodies or ashes' (Blackstone 1979:429).

[6] However, the common law provided some criminal and civil law protections to dead bodies.

[7] *Doodeward* v. *Spence* [1908] 6 CLR 406; *Dobson* v. *North Tyneside Health Authority* [1997] 1 WLR. 596; *R* v. *Kelly* [1988] 3 All ER 741; *Yearworth* v. *North Bristol NHS Trust* [2009] EWCA Civ 37.

[8] For instance, *Phillips* v. *Montreal General Hospital* [1908] *La Revue Legale* 159; Canada, LRC, supra, note 8. Also, in *Parpalaix* v. *CECOS* Gaz. Pal. 1984.2e sem.jur.560, a French court expressly refused to recognize the existence of property rights in stored sperm samples asserted by the wife of the deceased depositor; the property claims were intended to facilitate the claimant's efforts at posthumous procreation. However, the

In contemporary discourse on human tissues, the no-property rule serves as a rallying ground for those opposed to the objectification and commodification of the human body through modern biomedical technologies. It also expresses our shared sentiments and reverential attitude towards dead bodies and parts of a human body, an attitude reflected in the observation of the court in *Louisville & NR Co.* v. *Wilson* that 'the law relating to this mystery of what death leaves behind cannot be precisely brought within the letter of all rules regarding corn, lumber, and pig-iron.'[9]

Nevertheless, the no-property rule (and its associated traditional scholarship) is open to serious objections. It was developed in a period of relative scientific underdevelopment and hardly fits easily with recent technological innovations utilizing the human body and body parts (Nuffield Council on Bioethics 1995). Willes anticipated this objection in his thoughtful observation that in 'modern times the requirements of science are larger than formerly, and when they are so extensive it seems to me that we ought not to entertain any prejudice against the obtaining of dead bodies for the laudable purpose of dissection but we ought to look at the matter with a view to utility'.[10] Despite this early judicial reservation, the rule continued unabated. In sum, however, the no-property rule underpins the traditional analytical framework for cadavers and human issues. This traditional model of scholarship on human tissues is marked by its unidimensionality; its approach to questions of rights over the human body is often predictably inflexible. For instance, except for few cases falling within the narrow exceptions to the no-property rule, the traditional approach usually returns a verdict against the existence of property interests in corpses and most parts of the human body.

9.3 Critical approach

In contrast to the traditional analytical framework, the critical approach adopts a more robust theoretical and philosophical attitude towards questions of rights over the human body and body tissues (Davies and Naffine 2001). Consequently, the critical approach disavows the traditional proposition that cadavers and human tissues are automatically excluded from the realm of property law (ibid.). Instead of an a priori or talismanic application of the no-property rule, supporters of critical scholarship consider whether on conceptual, policy, and pragmatic grounds the human body should be admitted to the category of property. Recent writings on property theory, which consider the possibility of characterizing the human body and body parts as property, seem to vindicate the critical framework (Munzer 1990; Radin 1993). As noted above, the critical approach proceeds simultaneously from three viewpoints: theory, policy, and pragmatism. The analysis below turns on these tripartite dimensions of the critical framework for human tissues.

9.3.1 Theory

Arguably, there is nothing conceptually or theoretically objectionable about the categorization of the human body and tissues as property (Price 2007:119). Presumably, the 'non-propertization' of cadavers and most parts of the human body is not due to the failings of property theory but is based on the requirements of religion, statutory prohibitions, and social and governmental

court ordered the release of the sperm samples to the claimant on non-property principles (based on the terms of deposit and storage). However, in the Scottish (civil law) case of *H.M. Advocate* v. *Dewar* [1945] SLT 114 at 116, the High Court of Justiciary suggested (*obiter*) that, until burial, a corpse was capable of being stolen. This implies that property rights can exist in an unburied cadaver.

[9] *Louisville & NR Co.* v. *Wilson* [1905], 51 SE 24 at 25.

[10] *R* v. *Feist* [1858] 169 *Eng. Rep.* 1132 at 1135.

policies (Campbell 2009:11–26).[11] Major theories of property, for instance, could reasonably apply to the human body and body tissues. For example, the reified theory of property conceptualizes property as 'things' or physical entities. In other words, property rights are reified as physical things or objects of property (Litowitz 2000:401). Complaints that 'propertization' of human tissues objectifies the human body resonate with the reified theory of property. Reification also accords with the way in which lay persons commonly understand property (as 'things'). While reification of property carries the advantage of certainty and simplicity, it risks excluding commonly accepted forms of intangible property, such as intellectual property. There can be no doubt that the reified theory of property recognizes cadavers and visible body parts as property. Cadavers and human tissues separated from their sources have a physical presence that qualifies them as 'things' under the reified theory. *Roche* v. *Douglas* illustrates an interesting application of the reified theory to human tissues. In that case, Master Sanderson of the Supreme Court of Western Australia lamented that 'it defies reason to not regard tissue samples as property. Such samples have a real physical presence...There is no purpose to be served in ignoring physical reality'[12]. Therefore the reified theory potentially applies to biomedical inventions and reproductive materials that take a physical form, including gametes and cell lines. Similarly, disputes over frozen embryos, pre-embryos, and sperm illustrate further applications of the reified theory to human tissues.[13]

Another approach to property theory is to consider property as a bundle of rights. Thus property is not a 'thing' but the right exercisable over a thing. Therefore within the cognomen of property there are bundles of rights, powers, privileges, duties, and liabilities. A rights-based approach to property is quintessentially Hohfeldian; it employs jural relations to analyse normative concepts (Hohfeld 1911:16, 1917:710). Inherently, the bundle of rights approach fragments and disaggregates property. For instance, the property in my car is not so much the car as a physical thing as my rights over it, including my rights to exclusively use, sell, make a gift of, or even destroy the car. Legal ownership amounts to the fullest extent of these rights (Honoré 1961). Accordingly, every stick in the bundle of rights qualifies as property. For instance, if I am prohibited from destroying my car although I could make a gift of it to my friend, the car would still qualify as my property.[14] This is not surprising since property rights are not absolute in nature. I cannot turn my house into a brothel simply because I am the owner. However, that disability does not extirpate my property in the house.[15] In *National Provincial Bank* v. *Ainsworth*, however, Lord Wilberforce observed that before a right could properly qualify as property, 'it must be definable, identifiable by third parties, capable in its nature of assumption by third parties, and have some degree of permanence or stability'.[16] This implies that not every stick in the bundle would qualify as property; for though I might have an exclusive right to possess the dead body of my father for burial, I cannot make personal use of the corpse or sell it to third parties. Thus my possessory right over the corpse does not satisfy *Ainsworth's* criteria. However, it is more likely that Lord Wilberforce was referring to full legal ownership, the greatest accumulation of sticks in the bundle of property rights. As with the reified theory of property, the bundle of rights is potentially applicable to various aspects of human tissues and rights exercisable over them.

[11] For instance, the Human Tissue Act 2004 prohibits commercial transactions in body tissues.

[12] *Roche* v. *Douglas* [2000] WASC 146.

[13] For instance, *Yearworth*, supra, note 16; *Hecht* v. *Superior Court*, 20 Cal.Rptr. 2d. 275 (Ct. App. 19993); *Davis* v. *Davis*, 842 SW 2d 588 (Tenn. 1992).

[14] Compare *Blade* v. *Higgs* [1865] 11 ER 1471 (per Lord Westbury LC).

[15] For similar analysis: *First Victoria National Bank* v. *United States* [1980], 620 F. 2d 1096 at 1104.

[16] *National Provincial Bank* v. *Ainsworth* [1965] AC 1175 at 1247–8 (HL).

Arguably, the protection given to possessory rights over cadavers,[17] cultured human cells,[18] and stored human biological samples[19] illustrates the bundle of rights theory. Further illustration includes the protection given to users of excised tissues (against sources of tissues) and those who have turned human tissues into useful biomedical inventions.[20]

9.3.2 **Policy**

Satisfied that there are no insurmountable theoretical impediments to the recognition of human tissues as property, the critical framework proceeds to ascertain whether there are policy factors that weigh against the recognition of property in human tissues. Policy arguments for and against the recognition of property rights in excised human tissues are best illustrated by the case of *Moore* v. *Regents of the University of California* (ibid.) (see also Chapter 3, section 3.5). In that case, Moore's tissues were surreptitiously harvested in the course of a medically indicated procedure for hairy cell leukaemia. While the court upheld Moore's claim on informed consent, the majority struck out his conversion claim because of its adverse impact on biotechnology. The majority's policy decision resonates with the traditional framework for human tissues.[21] However, it is contradictory to deny Moore property rights in his excised tissues only to bestow the same property rights on those who unlawfully obtained his tissues. The preference for users over sources of tissues appears illogical. The majority's policy analysis is also difficult to justify. While a protective stance towards biotechnology might be economically and socially desirable, there is no empirical evidence that recognition of property rights in tissue sources would wreck biotechnology industries. Moreover, policy arguments can equally be made for tissue sources based on the need to protect their privacy, and the use of their tissues, through an enduring property right (Laurie 2002:299–304).

9.3.3 **Pragmatism**

Finally, critical scholarship justifies property rights in human tissues on pragmatic grounds. For instance, proponents of the critical framework argue that since the objectification and commodification of human tissues is already a social and economic fact, attention should rather focus on ways of regulating the emerging market (Mahoney 2000:163). It is undeniable that the market for human tissues has increased substantially due to the growth of the biotechnology industry. Products of the human body, such as hair, fingernails, toenails, bone marrow, and blood, have long been exchanged in commercial transactions, although in the case of blood payment was regarded (in some jurisdictions) as being made for service rather than for the blood itself, a sort of verbal detoxification.[22] There are also instances of current markets for stem cells harvested from fresh embryos (Munro 2001) and an antique auction for human tissues from a decommissioned anatomy laboratory (Sokoloff 2001). The existence of a black market for kidneys and other transplantable organs is also commonly acknowledged (Goodwin 2006). While some of the transactions above are unlawful (by statutes) in some jurisdictions, Justice Cowen of the US Court of

---

[17] *William* v. *William* [1882] 20 WLR. 659 (Ch. D).

[18] *US* v. *Arora*, 860 F. Supp. 1091 (D. Md. 1994).

[19] *Washington University* v. *Catalona* 437 F.Supp.2d 985 (Dist. Ct. Missouri, 2006).

[20] *Moore* v. *Regents of the University of California*, 793 P. 2d 479 9 (Cal. 1990); *Greenberg* v. *Miami Children's Hospital Res. Inst.* 264 F.Supp.2d 1064 (2003).

[21] For a criticism of this judgement and suggestions for using equitable principles to secure compensation for tissue sources see Hoppe 2007:199.

[22] *Perlmutter* v. *Beth David Hospital*, 123 NE 2d 792 (NY 1954); contrast *Green* v. *Commissioner* [1980], 74 TC. 1229.

Appeals recognized, in *Onyeanusi* v. *Pan Am*, that 'the very existence of these state laws indicates that there would be a market for human remains in the absence of government intervention'.[23] Thus, critics accuse the traditional analytical framework of paying insufficient attention to the entrepreneurial and market contexts of medical research and the corporatization of medical practice (Perry 1999).

9.4 **Modern approach**

In some ways, the modern approach to human tissues mirrors realist jurisprudence (Llewellyn 2008). Thus the modern analytical framework engages the practice of judges and adjudicators with a view to finding some discernible and predictable patterns of analysis in relation to human tissues. Therefore, at a general level, the modern framework posits that the judicial perception of human tissues is marked by a discriminatory analysis. In other words, when judges are called upon to determine the appropriate legal protection for human tissues they neither deny nor affirm the existence of property status for human tissues, although the traditional analytical model might provide a starting point for the judicial analysis. Instead, judges pay particular attention to the specific human tissue involved, whether or not it is excised from the body, whether the source is living or dead, whether the tissue was archived or subject to work and skill, whether the tissue was meant for research or exhibition, and whether the tissue is regenerative or non-regenerative. Great leeway exists for permutations of the different combination of circumstances in which judicial adjudication over human tissues can take place. Detailed analysis in that regard is beyond the scope of this chapter. Suffice it to say that in most common law jurisdictions judges are unlikely to recognize property rights over whole cadavers and living persons (Rao 2000). However, a different attitude applies to excised human tissues (from both cadavers and living sources). But this, in turn, depends on whether legal claims were made by sources or users and whether the tissues are regenerative or non-regenerative (and so on as above). Therefore the analysis below attempts to highlight the judicial treatment of various aspects of human tissues.

9.4.1 **Non-regenerative transplantable organs**

As regards non-regenerative transplantable organs, judges tend to be reluctant to recognize property rights in human tissues. However, non-property protection exposes an organ, such as a kidney, awaiting transplantation to unauthorized expropriation or destruction by third parties. Consider the case of an excised kidney donated to a donee and awaiting lodgement in the body of the intended recipient.[24] If the kidney is deliberately or negligently redirected to a third-party recipient, a wrong has been committed, but it is not clear that legal remedies are available to the donor and the donee (Nwabueze 2008). In the absence of statutory and criminal law considerations, significant remedial problems arise. In the case of deliberate (or intentional) redirection or misdirection, the action in *Wilkinson* v. *Downton*[25] might apply, but its utility is undermined by current controversy regarding the principle of the case; it is not yet clear whether *Wilkinson* actually established a cause of action for intentional tort. There are suggestions that *Wilkinson* established only a cause of action for negligence.[26]

---

[23] *Onyeanusi* v. *Pan Am*, 952 F 2d 788 at 792 (3d Cir. 1992).

[24] This assumes that the legal system allows directed organ donation, such as in the USA or for live transplants under the UK Human Tissue Act 2004.

[25] *Wilkinson* v. *Downton* [1897] 2 QB 57.

[26] For instance: *Page* v. *Smith* [1996] AC 155 at 177 (Lord Jauncey of Tullichettle); *Wainright* v. *Home Office* [2004] 2 AC 406 at 424 (Lord Hoffman). For a thorough exploration of this controversy see Nwabueze (2007a:191–6) and also Fleming (1998).

A claim in negligence, on the other hand, suffers from problems relating to causation and proof of damage. Arguably, battery claims are not applicable to separated body parts or organs. In any case, the initial consent to the harvesting of the organ is likely to defeat an allegation of battery. Consent-based actions might not be of much avail since consent does not confer continuing control over human tissues (Laurie 2002; Price 2003). An action in unjust enrichment is unlikely to succeed unless the misdirection was for the benefit of the wrongdoer, for instance to benefit a relative of the transplant surgeon. A claim in privacy is probably unavailing. The complaint is that an organ donated for the benefit of a particular recipient is lost through its misdirection, not that the donor or donee's privacy is infringed. Similarly, a claim in contract is hardly relevant. The misdirection is tortious but not contractual. Moreover, the statutory regulation of organs based on the principles of altruism (in most jurisdictions) militates against importing contractual principles into the context of organ transplantation. The above outline seems to leave property-based actions as the most opportune for the claimant in the hypothetical case under consideration. Interestingly, Matthews (1983) observed that where non-renewable organs are removed from the body 'one would have thought that as with blood, hair and the like that person had the first and best right to possession, though presumably one might transfer that right, as blood donors transfer it to a hospital or blood bank'; accordingly, he concluded that 'parts of the body once removed should be regarded as the "property", at least in a possessory sense, of the person from whom taken' (ibid.:227). Unfortunately, the courts have not generally treated kidneys and other non-renewable organs as property. The *Colavito* line of cases is illustrative.[27] There, the court held that no property right existed in a misdirected, but histo-incompatible, kidney.

9.4.2 **Human tissues for research**

Where human tissues are destined for research or employed in the research process, the courts are more likely (probably on policy grounds) to deny the property rights of sources as against those of the users. This judicial inclination is evident in *Moore* and *Greenberg* above. A more recent example is *Washington University* v. *Catalona*,[28] where Dr Catalona, an established cancer researcher at Washington University, moved to Northwestern University and sought to take with him some (cancer research) tissues from Washington University's biorepository. Tissues in the biorepository came from many sources, including participants in cancer research conducted by Dr Catalona. Before moving to Northwestern University, Dr Catalona got some of his research participants to sign documents which purported to mandate Washington University to release their tissues to Dr Catalona or Northwestern University. The claim was resisted by Washington University. Interestingly, all the parties in the case based their claims on property. As Limbaugh, the Senior District Judge realized, the 'sole issue determinative...of this lawsuit is the issue of ownership' (ibid.:994).

Limbaugh concluded that Washington University owned all the biological materials in its biorepository and that 'neither Dr William Catalona nor any research participant...has any ownership or proprietary interest in the biological samples housed in the GU Repository' (ibid.:1002). Limbaugh's judgement was affirmed by the Eight Circuit (US Court of Appeals)[29] and a request for further appeal to the US Supreme Court was denied.[30]

---

[27] *Colavito* v. *New York Organ Network (No. 1)*, 356 F. Supp. 2d 237 (EDNY 2005); *Colavito* v. *New York Organ Donor Network, No 2*, 438 F.3d 214 (2006); *Colavito* v. *New York Organ Donor Network (No. 3)*, 8 NY 3d 43 (2006, CANY); *Colavito* v. *New York Organ Donor Network (No. 4)*, 2007 WL 1462399, 1 (2nd Cir. N.Y). Detailed examination of the *Colavito* line of cases is provided elsewhere (Nwabueze 2008).

[28] *Washington University* v. *Catalona* 437 F.Supp.2d 985 (Dist. Ct. Missouri, 2006).

[29] *Washington University* v. *Catalona*, 490 F.3d 667 (8th Cir. 2007).

[30] *Catalona* v. *Washington University*, 128 S.Ct. 1122 (S. Ct. 2008).

9.4.3 **Human tissues subject to work and skill**

As noted earlier, most common law courts recognize that transformative work on a cadaver or human tissue could qualify it as property. This trend started in Australia with *Doodeward* v. *Spence*.[31] More recent common law cases have embraced the *Doodeward* exception.[32] For instance, in *AB* v. *Leeds Teaching Hospital NHS*, Gage observed that the 'principle that part of a body may acquire the character of property which can be the subject of rights of possession and ownership is now part of our law'.[33] The cases above show that the work and skill exception operates in favour of the service provider; in practice, this translates to users (biotechnology industries) rather than sources of tissues.

However, it is not clear what quantum of work and skill qualifies for the exception. In *AB* v. *Leeds Teaching Hospital NHS*, Gage was careful to explain the enormous amount of labour and skill required to prepare tissue blocks and slides. This implies that a lesser amount of work and skill would not trigger the exception, yet the foundational case of *Doodeward* involved such insignificant work—a double-headed fetus was merely preserved in a bottle of spirits. Nevertheless, the majority of the High Court of Australia recognized the double-headed fetus as property. In contrast, a similar preservation technique in *Dobson* v. *North Tynside HA* failed to trigger the *Doodeward* exception.[34] Thus uncertainty still exists as to the quantum of work and skill that qualifies for the exception.

9.4.4 **Products of the body and body wastes**

As regards body wastes and regenerative products of the body, the courts generally accept the existence of property rights in human tissues. For instance, the courts have adopted property analysis in relation to human faeces, hair, fingernails, and toenails.[35] A similar judicial attitude is discernible for blood[36] and urine.[37] While reproductive materials are specifically regulated by legislation in some jurisdictions,[38] the courts have generally accepted the existence of property rights in sperm.[39] Recent (unreported) decisions in the USA have confirmed the status of sperm as property,[40] in line with the foundational case of *Hecht* v. *Superior Court*.[41] In England, *Yearworth* v. *North Bristol NHS Trust* reflects the developments above.[42] Australia equally recognizes the

---

31 *Doodeward* v. *Spence* [1908] 6 CLR 406.

32 *R* v. *Kelly* [1998] 3 All ER 740 (CA).

33 *AB* v. *Leeds Teaching Hospital NHS* [2005] 2 WLR 358 at 394.

34 *Dobson* v. *North Tynside H.A.* [1997] 1 WLR 596.

35 *Venner* v. *Maryland* [1976], 354 A.2d 483; *R* v. *Herbert* [1961] JPLGR 12.

36 *R* v. *Rothery* [1976] RTR 550; *Carter* v. *Inter-Faith Hosp. of Queens*, 304 NYS. 2d 97 (App. Div. 1969) and other cases cited in footnote 22.

37 *R* v. *Welsh* [1974] RTR. 478.

38 Like the UK Human Fertilisation and Embryology Act 1990.

39 Despite the decision of a French Court in *Parpalaix* v. *CECOS*, supra, note 8, to the contrary.

40 CBC News, *Mother can have dead son's sperm harvested, Texas judge rules*, 8 April 2009. Available at: http://www.cbc.ca/health/story/2009/04/08/texas-sperm-mom008.html (accessed 15 April 2009); Associated Press, *Judge orders collecting of dead son's sperm*, 8 April 2009. Available at: http://www.chron.com/disp/story.mpl/headline/metro/6364234.html (accessed 15 April 2009); BBC News, *US woman gets dead fiancé's sperm*. Available at: http://newsvote.bbc.co.uk/mpapps/pagetools/print/news.bbc.co.uk/I/hi/world/america (accessed 28 April 2009).

41 *Hecht* v. *Superior Court* [1993], 20 Cal. Rptr. 2d 275.

42 *Yearworth* v. *North Bristol NHS Trust* [2009] All ER (D) 33.

existence of property rights in sperm.[43] However, it is unlikely that a pure property analysis would be adopted for human embryos.[44]

9.4.5 **Genetic information**

In some circumstances, such as genetic research, human tissues may be needed only for the information they contain. A person's gene is a significant source of biological, pathological, and ancestral information relating to that person. Thus genetic information is private, personal, and intimate. Unauthorized use of a person's genetic information could cause enormous psychological, social, and pecuniary damage. Consider, for instance, the use of a person's blood to determine his/her HIV status when only a diabetic test was authorized. A positive HIV test result might entail stigmatization or ostracism for the victim, as well as discrimination in employment and insurance. However, it is not clear that the victim of the above test could make property claims even though claims in privacy and consent might be available.[45] Property claims are critically important because they provide a source with ongoing control powers over his/her genetic information (Australian Law Reform Commission 2003:534). But the question is whether a person's genetic information should be recognized as property.[46]

The courts treat commercially valuable trade secrets, genetic or otherwise, as property.[47] Similarly, confidential (non-trade secret) information is, in practice, treated as property, although judicial analysis of confidence often proceeds on the basis of privacy rather than property.[48] But in other contexts common law courts seem unwilling to recognize information as property.[49] However, in *Boardman* v. *Phipps* the House of Lords held by a majority of three to two that non-genetic information would be recognized as property.[50] Thus the courts generally appear open to recognizing information as property (Weinrib 1988; Barrad 1993). This inclination is stronger in the area of intellectual property law where courts have generally confirmed the availability of patent protection for isolated genes and DNA sequences.[51] This means that courts generally recognize genes and genetic information from human tissues as property.

9.5 **Convergence of approaches: non-proprietary remedies**

Non-proprietary remedies provide a point of convergence for all analytical frameworks on human tissues. This convergence revolves around negligence, privacy, unjust enrichment, constitutional law, and human rights actions. In relation to human tissues, these causes of action apply with varying degrees of success. Generally speaking, however, non-proprietary forms of action evince

---

43 For instance, *Re Denman* [2004] QSC 070.

44 *Davis* v. *Davis* [1992], 842 SW 2d 588; *Kass* v. *Kass*, 696 NE 2d 174 (NY 1998); *Janicki* v. *Hospital of St Raphael* [1999], 744 A 2d 963.

45 For unsuccessful application of consent and negligence-based frameworks to unauthorized blood testing, see: *Doe* v. *Dyer-Goode*, 398 Pa. Super. 151 (Pennsylvania Sup. Ct. 1989); *Hecht* v. *Kaplan*, 221 AD 2d 100 (Sup. Ct. New York, 1996).

46 For detailed analysis of this issue, see Nwabueze (2007a:175–88).

47 *Pioneer Hi-Breed International* v. *Holden Seeds*, 35 F 3d 1226 (8th Cir. 1994).

48 *Campbell* v. *MGN Ltd* [2004] 2 AC 457 (HL).

49 *Oxford* v. *Moss* [1979] 68 Cr. App. R. 183; *R* v. *Stewart* [1988] *NR* 171 (SCC); *R* v. *Department of Health, ex parte Source Informatics Ltd* [2000] 1 All ER 786 (CA).

50 *Boardman* v. *Phipps* [1967] 2 AC 46. For the majority, on the issue of whether information is property: Lords Hodson, Guest, and Dilhorne.

51 For instance, *Relaxin Case* [1995] EPOR. 541; *Greenberg* v. *Miami Children's Hospital Res. Inst.*, 264 F. Supp. 2d 1064 (S.D. Fla. 2003).

significant weaknesses compared with property-based actions (Nwabueze 2007b). While all the analytical models explored above accept the regulation of human tissues through non-proprietary frameworks, the traditional framework diverges by its insistence on the exclusivity of non-proprietary remedies. In contrast, the critical and, to some extent, modern approaches suggest that while non-proprietary frameworks should be employed when necessary, they are by no means exclusive. Accordingly, the critical and modern analytical models support the use of property together with other legal categories.

9.6 **Conclusion**

This chapter suggests that the legal status of human tissues is still evolving. The question of whether a human tissue deserves protection or what legal category underpins that protection often depends on one or more paradigms generally applied to the analysis of human tissues. Accordingly, this chapter set out to identify and explore the tripartite frameworks which usually animate the analysis of human tissues. While no preference is given to any of the tripartite frameworks, this chapter suggests a trend towards a more inclusive use of all analytical models, in contrast with the monolithic approach induced by the no-property rule.

References

Australian Law Reform Commission (2003). *Essentially yours: the protection of human genetic information in Australia*. Australia: ALRC.

Barrad, C.M. (1993). Genetic information and property theory. *Northwestern University Law Review*, **87**, 1037–86.

Blackstone, W. (1979). *Commentaries on the laws of England*. Chicago: University of Chicago Press.

Campbell, A.V. (2009). *The body in bioethics*. London: Routledge–Cavendish.

Davies, M. and Naffine, N. (2001). *Are persons property? Legal debates about property and personality*. Aldershot: Ashgate.

Fleming, J.J. (1998). *The law of torts* (9th edn.), pp.31–41. Sydney: LBC Information Services.

Goodwin, M. (2006). *Black markets: the supply and demand of body parts*. New York: Cambridge University Press.

Gordon-Grube, K. (1988). Anthropophagy in post-Renaissance Europe: the tradition of medical cannibalism. *American Anthropology*, **90**, 405–9.

Hohfeld, W.N. (1911). Some fundamental legal conceptions as applied in judicial reasoning. *Yale Law Journal*, **23**, 16–59.

Hohfeld, W.N. (1917). Some fundamental legal conceptions as applied in judicial reasoning. *Yale Law Journal*, **26**, 710–70

Honoré, T (1961). Ownership. In A.G. Guest (ed.), *Oxford essays in jurisprudence*, pp.107, 118–19. Oxford: Oxford University Press

Hoppe, N. (2007). Out of touch: from corporeal to incorporeal, or Moore revisited. In C. Lenk, N. Hoppe, and R. Andorno (eds), *Ethics and law of intellectual property: current problems in politics, science and technology*. Aldershot: Ashgate

Laurie, G. (2002). *Genetic privacy: a challenge to medico-legal norms*. Cambridge: Cambridge University Press

Law Reform Commission of Canada (1992). *Procurement and transfer of human tissue and organs. Working Paper 66*. Ottawa: Law Reform Commission of Canada.

Litowitz, D. (2000). Reification in law and legal theory. *Southern California Interdisciplinary Law Journal*, **9.**

Llewellyn, K.N. (2008). *The bramble bush*. New York: Oxford University Press.

Mahoney, J.D. (2000). The market for human tissue. *Virginia Law Review*, **86**, 163–223.

Matthews, P. (1983). Whose body? People as property. *Current Legal Problems*, **36**, 193–239.

Munro, M. (2001). A vision of spare parts. *National Post (Canada)*, 29 March 2001, A15.

Munzer, S.R. (1990). *A theory of property*. New York: Cambridge University Press.

Nuffield Council on Bioethics (1995). *Human tissue: ethical and legal issues*. London: Nuffield Council on Bioethics.

Nwabueze, R.N. (2002). Spiritualizing in the godless temple of biotechnology: ontological and statutory approaches to dead bodies in Nigeria, England, and the USA. *Manitoba Law Journal*, **29**, 171–219.

Nwabueze, R.N. (2007a). *Biotechnology and the challenge of property; property rights in dead bodies, body parts and genetic information*. Aldershot: Ashgate.

Nwabueze, R.N. (2007b). Interference with dead bodies and body parts: a separate cause of action in tort? *Tort Law Review*, **15**, 63–6.

Nwabueze, R.N. (2008). Donated organs, property rights and the remedial quagmire. *Medical Law Review*, **16**, 201–4.

Perry, M.M. (1999). Fragmented bodies, legal privilege, and commodification in science and medicine. *Maine Law Review*, **51**, 169–210

Price, D. (2003). From Cosmos and Damien to Van Velzen: the human tissue saga continues. *Medical Law Review*, **11**, 1–47.

Price, D. (2007). Property, harm and the corpse. In B. Brooks-Gordon, F. Ebtehaj, J. Herring, M. Johnson, and M. Richards (eds). *Death rites and rights*. Oxford: Hart.

Radin, J. (1993). *Reinterpreting property*. Chicago, IL: University of Chicago Press.

Rao, R. (2000). Property, privacy, and the human body. *Boston University Law Review*, **80**, 359–460.

Richardson, R. (2006). Human dissection and organ donation: a historical and social background. *Mortality*, **11**, 151–65.

Scott, R. (1981). *The body as property*. New York: Viking Press.

Sokoloff, H. (2001). Human tissue on sale at auction. *National Post (Canada)*, 16 April 2001, A4.

Waldby, C. and Robert, M. (2006). *Tissue economies: blood, organs and cell lines in late capitalism*. Durham: Duke University Press.

Weinrib, A.S. (1988). Information and property. *University of Toronto Law Journal*, **38**, 117–50.

Chapter 10

Research with human biological material and personal data in biobanks: legal and regulatory framework in Switzerland

Bianka S. Dörr

10.1 Introduction

As a result of the successful sequencing of the human genome and the rapid progress in the field of molecular genetics and biotechnological analytical processes, a new category of sample collections has emerged since the middle of the 1990s, the so-called biobanks. Biobanks are private or public institutions which gather and store samples of human bodily substances (cells, blood, organs, tissue) or their constituents (serum, DNA) in order to correlate them with donor-related data (genes, phenotype, disease, lifestyle) and provide them for scientific research purposes. Thus the characteristic of a biobank is a combination of material and information (Dörr 2007:446). However, as outlined in previous chapters (see Part I of this volume), the use of biological material in the context of scientific research and biomedical applications, as well as the creation and operation of biobanks worldwide, raises a number of questions with regard to collecting, storing, using, and transferring samples and data. In order to face these challenges, the Swiss legislator has been working on the implementation of a federal act regarding research involving humans (*Humanforschungsgesetz*) and recently published a draft bill and the corresponding Federal Council Dispatch (*Botschaft*). In addition, in May 2006 the Swiss Academy of Medical Sciences (SAMS) published guidelines and recommendations regarding biobanks entitled *Biobanks: acquisition, preservation and use of human biological material* (SAMS 2006). In November 2008, these guidelines and recommendations were amended by another set of recommendations entitled *The use of dead bodies and parts thereof for medical research, teaching and education* (SAMS 2008).

These regulatory activities in Switzerland regarding biobanks as well as research with human materials and personal data are described and critically evaluated in this chapter,. As the Swiss legislative and regulatory activities try to ensure that the participating individuals are protected, while at the same time providing optimal research conditions for the use of the material and data collected, they may serve as a model for other legislators and help to foster international discussion in this field.

10.2 Biobanks and networks in Switzerland

The exact number of biobanks in Switzerland is not known. It is estimated that there are a large number of small privately and publicly run biobanks containing a few hundred to a few thousand stored samples. There are no large biobanks with population-related data like the UK Biobank (see Chapter 5, section 5.2).

10.2.1 The Biobank Suisse Foundation

Recently, the Biobank Suisse Foundation was set up as a non-profit foundation in order to establish a Swiss network of biobanks. The aim of the network is to promote cooperation between university and non-university research institutes, as well as other institutions working in life sciences and bioinformatics, and to encourage collaboration between scientists and clinical practitioners as well as providing researchers with high-quality materials for their studies[1].

10.2.2 SAPALDIA

The SAPALDIA biobank was established within the scope of the SAPALDIA research project (Swiss Study on Air Pollution and Lung Diseases in Adults), which started in 1991. This population-based cohort study is examining eight selected localities in Switzerland with an air pollution problem and analysing the relation between air quality and the incidence and severity of contracted respiratory diseases in adults[2].

10.3 Research with biological material and personal data in biobanks

10.3.1 Present legal framework at federal level

Although the majority of the Swiss cantons have enacted at least some regulations on medical research involving humans at a cantonal level, currently there is no uniform legislation or regulation of research involving humans or human biological material and personal data at federal level. Thus, many scientific projects encounter considerable legal uncertainty when starting their collections and using human biological material in research (Dörr 2007:466). So far, only the use of stem cells, germ-line cells, and embryos is regulated in the Federal Act on Medically Assisted Reproduction (*Fortpflanzungsmedizingesetz*) and the Federal Act on Research with Embryonic Stem Cells (*Stammzellenforschungsgesetz*). Article 20, paragraph 2, of the Federal Act on Human Genetic Analyses (*Bundesgesetz über genetische Untersuchungen beim Menschen*) allows genetic analyses using biological material extracted for other purposes than research, provided that the anonymity of the donors is guaranteed and the further use of the material has not been explicitly forbidden by the donors after they have been given appropriate information. The data obtained from these analyses are subject to the general data protection and professional secrecy regulations. However, neither the Federal Act on Data Protection nor Article 321 pp. of the Swiss Penal Code, which rules on breaches of professional secrecy, contain any special provisions for the use of data related to human biological material.

Based on the 'Plattner Motion' of 1 December 1998 (Ständerat 1998), the Swiss legislator is currently working on a comprehensive act on research involving humans (*Humanforschungsgesetz*) which aims, inter alia, to regulate research using samples of human biological material and health-related personal data. Work on this act was started in the middle of 2000 but suffered prolonged interruption in autumn 2001, when the Federal Council enshrined the ex ante regulation of research on surplus human embryos and human embryo stem cells in a separate act (Botschaft 2009:9). Work on the planned legislation on research involving humans was not resumed until Parliament passed the Act on Research with Embryonic Stem Cells on 19 December 2003. After releasing a preliminary draft proposal on research involving humans in February

[1] http://www.biobank-suisse.ch (accessed 28 January 2010).

[2] http://www.sapaldia.net (accessed 28 January 2010).

2006, the Federal Council started a four-month consultation procedure on the proposed legislation. This consultation procedure (*Vernehmlassungsverfahren*) is a special phase in the Swiss legislative process which primarily serves to obtain information on the factual accuracy, feasibility of implementation, and acceptance of federal proposals, particularly proposals that are subject to a referendum such as constitutional amendments and legislative bills. The interested parties (e.g. the cantons, political parties, communities, town authorities, and other associations) are invited by the Federal Council to submit their opinions on the proposals[3]. The reactions of the parties involved are subsequently evaluated and published.

A total of 153 statements were submitted to the consultation procedure. Two-thirds of them assessed the preliminary draft proposal positively or neutrally, while one-third gave it a negative appraisal, judging it as either too friendly or too hostile to research (Eidgenössisches Departement des Innern 2007:32; Botschaft 2009:30). During the last two and a half years the preliminary draft proposal has undergone a thorough revision. On 21 October 2009 the draft bill on research involving humans and the corresponding Federal Council Dispatch were presented to the Swiss Parliament for further consultation. Compared with the preliminary draft proposal, the draft bill has been considerably restructured and streamlined, and some technical or detailed provisions will be implemented as regulatory by-laws at a later date. In order to facilitate entering into international cooperation and international research involving humans, the draft bill has been amended to take account of the relevant international conventions, guidelines, and recommendations such as, inter alia, the Oviedo Convention and the Additional Protocol concerning Biomedical Research, the EU Tissues and Cells Directive (2004/23/EG), and the Declaration of Helsinki (1964/2008). (Botschaft 2009:23, 24, 26)

At the same time, a new constitutional provision on research involving humans was elaborated (Swiss Federal Constitution, art. 118b) which aims to grant the Confederation comprehensive competence to regulate in this field. On 25 September 2009 a Federal decree on the appropriate constitutional provision was passed by the Swiss Parliament (Bundesbeschluss 2009:6649). In a referendum (*Volksabstimmung*) on the new constitutional provision, which took place on 7 March 2010, the vast majority of Swiss voters agreed to its implementation.

The draft bill of the act on research involving humans defines its main objective as the protection of the dignity and personality of humans in research; additionally, health as part of the right of personality is explicitly mentioned, as it deserves particular importance in the context of research involving humans (art. 1 para. 1). Furthermore, the fundamental right of freedom of research is to be sustained and balanced with the right of personality (art. 1 para. 2 lit. a). Likewise, the planned act will help to ensure quality (art. 1 para. 2 lit. b) and transparency (art. 1 para. 2 lit. c) of research involving humans. The draft bill applies to research on human diseases as well as to research on the structure and function of the human body that will be carried out on living and dead persons, embryos and fetuses, biological material, and health-related personal data (art. 2 para. 1). However, research with anonymized biological material or with anonymously collected and anonymized health-related personal data is explicitly excluded from the scope of the draft bill (art. 2 para. 2). This is because research-related handling of these materials has been assessed using hazard analysis and found not to present any potential risks for the persons concerned (Botschaft 2009:38–9).

The revised provisions of the draft bill stipulate in a separate section those fundamental principles that are enshrined in national and international regulations and should be observed generally in any research: the primacy of the interests, health, and welfare of human beings over

---

[3] http://www.admin.ch (accessed 28 January 2010).

the interests of science and society (art. 4), the scientific relevance of the research projects (art. 5), non-discrimination (art. 6), informed consent (art. 7), right to information (art. 8), and prohibition of commercialization of the human body (art. 9).

The revised Chapter 2 sets out general conditions for research with persons and the taking of biological material and health-related personal data, for example informed consent and its withdrawal, protective measures for participating persons, and a liability scheme (art. 11–20). Additional conditions for on research with vulnerable persons are established in Chapter 3 (art. 21–30). Chapter 4 governs the further use of biological material and personal data (art. 31–34). Here, a conceptual distinction between genetic data and non-genetic health-related personal data is introduced. The notion *genetic data* comprises all information about the genome of a person obtained by a genetic test (art. 3, lit. g), and the term *non-genetic health-related personal data* covers all information about the health or illness of an identified or identifiable individual, except their genetic data (art. 3, lit. f). As the Swiss legislator estimates that, because of its clear information content, the risk of misuse of non-genetic health-related personal data is lower than that of the abuse of genetic data, research using such data is subject to reduced conditions (Botschaft 2009:39).

The proposed scheme for the further use of material and data for research is, of course, based on the traditional principles of informed consent. In addition, a new gradual regulation concept (*abgestuftes Regelungskonzept*) is provided, depending upon whether research with biological material or genetic data or non-genetic health-related personal data is concerned and whether those materials and data are presented in an uncoded, pseudonymized, or anonymized form. In the case of uncoded biological material and genetic personal data, further use of these materials and data is only allowed if the participants have consented to each research project. In these circumstances, obtaining general consent is clearly not permissible (art. 31 para. 1). However, general consent is regarded as being the rule if further use of pseudonymized biological material and genetic data is at issue (art. 31 para. 2). As laid down in art. 31, para. 3, the anonymization of biological material and genetic data for research purposes is allowed if the persons concerned have been given advance information and have not given any veto; here, consent is constructed as presumed consent. This information may, for example, form part of a hospital's patient information brochure (Botschaft 2009:77). General consent is also established as the rule for further use of uncoded non-genetic health-related personal data (art. 32 para. 1). Presumed consent (prior notice and no veto) is also regarded to be sufficient with respect to further use of pseudonymized non-genetic health-related personal data (art. 32 para. 2). However, if the requirements for consent and information cannot be met, biological material or health-related personal data may exceptionally be used for further research, inter alia, under the following conditions: it is impossible or extremely difficult to obtain consent, to inform about the right of withdrawal, or to expect this from the person concerned (art. 33 lit. a); no documented withdrawal is available (art. 33 lit. b); or the interest of research on the further use of the material and data prevails over the interests of the persons concerned (art. 33 lit. c). A small amount of material may be taken without consent from bodily substances removed during a post-mortem or transplantation and be anonymized for research purposes provided that there is no documented refusal by the deceased (art. 37).

Because of the strong objections that wer expressed in the consultation procedure regarding the preliminary draft proposal, the provisions regarding authorization or compulsory registration requirements for biobanks have been omitted from the draft bill (Botschaft 2009:39).[4] Similarly, to

[4] In this respect, the Swiss proposition on the regulation of human tissue research is more liberal than the French legal framework, which makes the authorization of biobanks obligatory (see Commin in this volume).

facilitate international research collaborations, the prohibition on exporting uncoded material and data for research purposes has been modified considerably compared with the preliminary draft proposal as the draft bill now stipulates that uncoded biological material or genetic data may be exported abroad for research if the persons concerned have given consent; general consent is regarded as being sufficient for pseudonymized material and data (art. 41) (for the issue of sample transfer, see Chapter13).

10.3.2 Medico-ethical guidelines and recommendations by the Swiss Academy of Medical Sciences

For many years, the Swiss Academy of Medical Sciences (SAMS), an independent foundation established in 1943 with its headquarters in Basle, has concentrated its activities on, inter alia, establishing guidelines and recommendations on topical medical issues as well as on clarifying ethical problems related to biomedical and technical developments and their effects on society. All these activities are primarily concerned with the Swiss environment but they are also strongly oriented towards international developments.[5] In Switzerland, the SAMS guidelines are valued due to their topicality and flexibility which allow a rapid reaction to changes (Rüetschi 2004:1222–3). Although the guidelines and recommendations published by SAMS are, strictly speaking, not legally binding (ibid.:1222), as a rule they are incorporated into the Statutes of the Swiss Medical Association and therefore they are binding on members of the Association. Furthermore, they have acquired importance indirectly because several cantons refer to them in their own regulations and they are regularly used as an orientation guide in court (ibid.:1222–3; Botschaft 2009:26). SAMS has recently published a *Compendium on research with humans* which aims to serve as a practical guideline by designing and assessing research projects involving humans (SAMS 2009).

The medico-ethical guidelines and recommendations *Biobanks: obtainment, preservation and use of human biological material* were published by the SAMS to serve as an orientation guide until the comprehensive federal regulation (the Act on Research with Humans) becomes effective (SAMS 2006:I). The guidelines include provisions to protect participants' dignity and personality as well as regulations to guarantee the quality and safety of biobanks. The content of the guidelines content is based on the premise that transparent rules will increase trust in research in the long term and therefore will have a positive influence on potential donors of biological material (Dittmann and Salathé 2006:1023). The guidelines are addressed to all operators and users of public and private biobanks, as well as other collections of human biological material, and are applicable to research, teaching, and education (SAMS 2006:II, art.2). However, they do not apply to the use of tissue material for individual diagnosis, therapeutic or forensic purposes, or quality control and safety, as long as this occurs within medical practice (ibid.).

10.3.2.1 Quality and safety management

Quality and safety management features prominently in Article 3 of the SAMS guidelines. A detailed schedule of provisions encompassing various demands regarding quality standards, data protection, and transfer of samples and data, as well as estabishing organizational rules, is stipulated.

10.3.2.2 Research with human biological material

Article 4 on research with human biological material comprises detailed regulations aimed at protecting the participant's dignity and privacy. Article 4.1 stipulates that any research project

[5] http://www.samw.ch (accessed 28 January 2010).

using human biological material that may directly affect the donor is subject to prior positive approval by the appropriate Ethics Committee for Clinical Trials. This applies in particular to research projects in which samples of human biological material are taken for research purposes, but it also applies to research projects using pseudonymized or uncoded samples and data.

With regard to the important issue of information and consent, the guidelines on biobanks provide the following guidance. As a general rule, donors must give their consent for removing, storing, and using samples of their biological material for research purposes; written consent preceded by written information must be available at the latest when the samples and data are stored in a biobank (SAMS 2006:II, art. 4.2, 4.3). The extent of the information given to the donor should be appropriate to the planned use of the samples and data (ibid.:II, art.4.2). Article 4.2 contains a list of items that are considered particularly relevant for donors which should be included in the written information. Among these are measures taken for the protection of personality, the period of storage of samples and data, the right of access to stored data, and the field of application.

As a rule, consent is given in a general form, allowing further use of samples and data for future research projects (ibid.:II, art.4.3). However, there is a standardized right to withdraw consent, which applies to pseudonymized samples and data and their future use (ibid.:II, art.4.6). In the event of a withdrawal of consent, samples must be destroyed. This does not apply to the results obtained with this material and their subsequent analyses (ibid.). Moreover, the biobank guidelines stress the right of the donor to be informed by the responsible physician of any diagnostically or therapeutically relevant result (right to know), as well as, to the contrary, the right to refuse such information (right not to know) (ibid.:II, art.4.7). Likewise, the transfer of samples and data, which is only possible in anonymized or coded form, requires prior consent from the donor (ibid.:II, art.4.8). Even biobanks that were in existence before the SAMS guidelines were established should try to fulfil the requirements and observe the principles from now on. As in these cases a donor's consent to the further use of the existing samples and data is often neither available nor a posteriori obtainable, research projects require, in addition to approval of the Ethics Commission for Clinical Trials, authorization from the Expert Commission on Professional Secrecy in Medical Research (ibid.:II, art.4.9).

10.3.2.3 Human tissue preparations in collections, exhibitions and museums

The SAMS guidelines on biobanks also include provisions concerning human tissue preparations in collections, exhibitions, and museums (ibid.:III).

10.3.2.4 SAMS recommendations

Finally, the SAMS guidelines make the following recommendations to the federal authorities: (1) creation of registers of public and private biobanks; 2) establishment of standards for education in laboratory work; (3) development of provisions for the accreditation of biobanks; (4) creation of an information and consent form to be signed by patients before their admission to a hospital (ibid.:IV).

10.4 **Discussion and outlook**

For some years, the number and size of biobanks containing human biological material and personal data have been increasing worldwide. At the same time, given the manifold ethical, legal, and social demands, the need for clarification with respect to research using these materials and data, the handling of the stored materials, and the creation and operation of biobanks is also increasing. For this reason, the creation of a coherent regulatory scheme for this research is becoming an increasingly important task for the legislature. The legislator faces the difficult task of appropriately balancing the interests of scientific knowledge, the law, and public welfare on the one hand, and protection of the personality of participants on the other, within the frame of existing legal provisions.

By examining whether the draft bill and the SAMS guidelines fit into the current Swiss legal framework and how the conflicting interests between personal protection and freedom of research are reconciled, the basic human rights of the Swiss Federal Constitution provide a benchmark for action: the right to protection of personality on one hand, and the right to freedom of research on the other hand.

10.4.1 Protection of personality

The constitutional protection of personality has different aspects. In addition to life, physical and mental integrity, and personal freedom, the development of a person, his/her private and family life, and his/her autonomy and self-determination must be ensured (Schweizer 2001:693). Disease-related and clinical research, which is one of the most common forms of research using humans, affects the participating individuals as it almost always constitutes interference with their bodies and their physical integrity. Every action which involves taking and also storage and use of biological materials must take place voluntarily and is subject to the autonomous decision of the person concerned, regardless of whether the material is obtained as part of a diagnostic or therapeutic action or specifically for research purposes (Haas, 2007:255; Büchler and Dörr 2008:397). In the context of biobanks, the participant's right of informed self-determination has particular importance. The autonomy of persons and the guarantee of their self-determined development in a free society, in relation to any handling of their personal data, form, as expressions of human dignity, the central point of the right of informed self-determination (Müller 1999:45; St Galler Kommentar 2008:art. 13 N 40). Individuals are not only protected against misuse of their personal data, but are also free to decide what value they attach to their own data and which data they wish to disclose to foreign, public, or private handling, and the purposes for which these data may be used (Mund 2005:86; Dörr 2007:462–3; St Galler Kommentar 2008:art. 13 N 38). As data are particularly worthy of protection, its processing requires the explicit informed consent of the individuals concerned. In addition, important elements of the right to informed self-determination form certain information rights and protection claims, such as the right to know everything that concerns one's own person (in particular, diagnoses and results of research) or the right not to know. The last protects a person's self-determined and autonomous decision to abstain from taking notice of his/her genetic information as well as the right not to be confronted with available knowledge about his/her genetic make-up (Mund 2005:39, 106; Dörr 2007:463–4). The right to information and access or the right to correction or deletion of erroneous data are also part of the right to informed self-determination (Müller 1999:47–8; Rhinow 2003:sect. 14 N 1280; St Galler Kommentar 2008:art. 13 N 44).

10.4.2 Freedom of academic research

The basic right to academic freedom is anchored in Article 20 of the Swiss Federal Constitution. The scope of the provision covers freedom of academic teaching as well as freedom of academic research (Schwander, 2002:91, 112). All activities of the researcher necessary for the collection of knowledge fall within the scope of protection of the freedom of academic research, which is of interest in the present context (ibid.:92, 113; Müller 1999:319–20; Kiener and Kälin 2007:240). To be precise, this covers the entire research process, consisting of the actual research activity and all necessary preparatory and follow-up work as well as publication and communication of the related results (Schwander 2002:92, 113ff.; Müller 1999:319). Thus researchers are not subject to state influence in their intellectual and methodological independence, and are protected in their quest for knowledge (Kiener and Kälin 2007:241). However, academic freedom is not guaranteed without limits; it may be limited in compliance with certain constitutional requirements for the protection of other legal interests (Schwander 2002:177; Kiener and Kälin 2007:75, 243–4). In particular, academic freedom is limited when the personality rights of third parties, public health, or the protection of the environment are affected (Rhinow 2003:sect.16 N 1516).

One possible way of balancing these basic rights lies in the design of the informed consent scheme. If, as favoured by the SAMS guidelines and the draft bill, a general consent scheme is selected which enables the donors of biological material to consent to unknown (further) use(s) of their samples and data for future research purposes at the time they consent, this is not an entirely unproblematic approach, at least from a Swiss legal perspective (for an alternative approach to informed consent, see Chapter 2). Under Swiss law, human bodily substances and data may only be used for those purposes of which the donor is aware and has agreed to (Haas 2007:200). Thus donors must be adequately informed, in advance, about the purpose, importance, threats, risks, and benefits of the collection, storage, and use of their biological materials. A particular level of certainty (*Bestimmtheit*) is always required as part of this condition (ibid.:287; Büchler and Dörr 2008:398). As the spectrum of future research is often not predictable at the time of removal or storage of bodily materials, giving of such an exhaustive consent contradicts not only the purpose limitation principle (*Zweckbindung*) intrinsic to each informed consent, but can also mean that donors have a significant loss of control over the handling of their material and data (ibid.:402–3). In addition, the right to (informed) self-determination can only be properly exercised if an appropriate level of information and knowledge is ensured; otherwise it is impossible for donors to understand the dimensions of their declaration. Moreover, the Swiss Civil Code (art. 27 para. 2) prevents persons from binding themselves to indefinite purposes for an indefinite period of time (Haas 2007:287; Büchler and Dörr 2008:402–3). Therefore general consent does not comply with said requirements.

When the draft bill and the SAMS guidelines on biobanks are assessed, it must be noted, first, that both documents are committed to anchoring and strengthening subjective provisions for the protection of dignity and personality of those individuals who are participating in research projects, or will be asked to do so, and at the same time to establishing protective objective measures, such as, inter alia, the examination of research projects by an ethics committee for research, the obligatory recording of approved research projects in a public register, the project's relevance to scientific research, or the establishment of provisions for quality and safety of biobanks (in the SAMS guidelines). Secondly, both documents provide greater legal certainty and transparency in research that, in turn, may help to enhance public trust in research with humans and in science. Nevertheless, when it comes to the question of a general consent, both legal documents are very research-friendly, as this form of consent represents the greatest courtesy available to researchers and the freedom of research.

References

Botschaft zum Bundesgesetz über die Forschung am Menschen [Federal Council Dispatch regarding the federal law on research involving humans] (2009). *Bundesblatt*, 8045.

Büchler, A. and Dörr, B.S. (2008). Medizinische Forschung an und mit menschlichen Körpersubstanzen. Verfügungsrechte über den Körper im Spannungsfeld von Persönlichkeitsrechten und Forschungsinteressen [Medical research on and with human body substances. Rights of disposal over the body between the poles of personality rights and research interests]. *Zeitschrift für Schweizerisches Recht*, **127**, 381–406.

Bundesbeschluss zu einem Verfassungsartikel über die Forschung am Menschen [Federal decree on a constitutional article on research involving humans] (2009). *Bundesblatt*, 6649.

Dittmann, V. and Salathé, M. (2006). Übergangsregelung für Biobanken bis zum Inkrafttreten des neuen Humanforschungsgesetzes [Interim regulation for biobanks until the implementation of a new human research law]. *Schweizerische Ärztezeitung*, **23**, 1022–3.

Dörr, B.S. (2007). Blut, Gewebe, Zellen, DNA und Daten—Biobanken im Spannungsfeld von Persönlichkeitsrechten und Forschungsvisionen [Blood, tissue, cells, DNA, and data—biobanks between the poles of personality rights and research visions]. In B.S. Dörr and M. Michel (eds) *Biomedizinrecht. Entwicklungen Perspektiven Herausforderungen*, pp. 443–75. Zürich: Dike.

Eidgenössisches Departement des Innern (2007). *Bericht über die Ergebnisse des Vernehmlassungsverfahrens zum Vorentwurf einer Verfassungsbestimmung und eines Bundesgesetzes über die Forschung am Menschen [Report on the results of the consultation procedure on the draft of a constitutional clause and a federal law on research involving humans]*. Available at: http://www.admin.ch/ch/d/gg/pc/documents/1266/Ergebnis.pdf (accessed 15 December 2009).

Haas, R (2007). *Die Einwilligung in eine Persönlichkeitsverletzung nach Art. 28 Abs. 2 ZGB [The consent to a violation of personality]*. Zürich: Schulthess.

Kiener, R. and Kälin, W. (2007). *Grundrechte [Fundamental rights]*. Bern: Stämpfli.

Morr, U. (2005). *Zulässigkeit von Biobanken aus verfassungsrechtlicher Sicht [The legitimacy of biobanks in constitutional law]*. Frankfurt am Main: Peter Lang.

Müller, J.P. (1999). *Grundrechte in der Schweiz [Fundamental rights in Switzerland]*. Bern: Stämpfli.

Mund, C. (2005). *Grundrechtsschutz und genetische Informationen [Protection of fundamental rights and genetic information]*. Basel: Helbing Lichtenhahn.

Rhinow, R. (2003). *Grundzüge des Schweizerischen Verfassungsrechts [Main features of the Swiss constitutional law]*. Basel: Helbing Lichtenhahn.

Rüetschi, D. (2004). Die Medizinisch-ethischen Richtlinien der SAMW aus juristischer Sicht [The medico-ethical guidelines of the SAMW from a juridical perspective]. *Schweizerische Ärztezeitung*, **23**, 1222–5.

SAMS (Swiss Academy of Medical Sciences) (2006). *Biobanken: Gewinnung, Aufbewahrung und Nutzung von menschlichem biologischem Material für Ausbildung und Forschung [Biobanks: acquisition, preservation and use of human biological material]*. Available at: http://www.samw.ch/de/Ethik/Richtlinien/Aktuell-gueltige-Richtlinien.html (accessed 15 December 2009).

(SAMS) Swiss Academy of Medical Sciences (2008). *Verwendung von Leichen und Leichenteilen in der medizinischen Forschung sowie Aus-, Weiter- und Fortbildung. [The use of dead bodies and parts thereof for medical research, teaching and education]*. Available at: http://www.samw.ch/de/Ethik/Richtlinien/Aktuell-gueltige-Richtlinien.html (accessed 15 December 2009).

(SAMS) Swiss Academy of Medical Sciences (2009). *Forschung mit Menschen [Compendium on research with humans]*. Available at: http://www.samw.ch/de/Publikationen/Leitfaden.html (accessed 15 December 2009).

Schwander, V. (2002). *Grundrecht der Wissenschaftsfreiheit im Spannungsfeld rechtlicher und gesellschaftlicher Entwicklungen [Fundamental right of the freedom of science between the poles of legal and societal developments]*. Bern: Stämpfli.

Schweizer, R.J. (2001). Verfassungsrechtlicher Persönlichkeitsschutz [Constitutional protection of personality] in D. Thürer, J.-F. Aubert, and J.P. Müller (eds). *Verfassungsrecht der Schweiz*, pp. 691–706. Zürich: Schulthess.

Ständerat (1998). *Motion 98.3543 (Plattner) vom 1. Dezember 1998, Schaffung eines Bundesgesetzes betreffend medizinische Forschung am Menschen [Motion 98.3543 of 1 December 1998: creation of a federal law concerning medical research on humans]*. Available at: http://www.parlament.ch/poly/Suchen_amtl_Bulletin/ce99/printemp/320.HTM (accessed 30 January 2010).

St. Galler Kommentar (2008). *Die schweizerische Bundesverfassung [The Swiss Federal Constitution]*. Zürich: Dike/Schulthess.

Chapter 11

Legal issues surrounding French research-focused biobanks

Virginie Commin

The number of research biobanks around the world has been increasing because of the health improvements that genetic tests and gene therapy promise to bring to the field of predictive medicine, and because of the improvements that individual genetic profiles might bring to the field of personalized medicine (Fletcher 2008). We can distinguish between two types of biobanks: '[those] based on biological specimens from patients or donors, and the population-based research biobanks that are based on biological samples from healthy and sick members of the general population' (Gottweis and Petersen 2008). Population-based research biobanks focus on the most common diseases (cardiovascular diseases, neurological diseases, etc.) and on monitoring the health of individuals relative to their lifestyle over the course of several years. These population-based research biobanks, to which anyone can donate samples, exist in the UK, Estonia, Iceland, Sweden, and China, among other places.[1] Other biobanks can be classified as disease-centred, i.e. dedicated to research on a specific type of disease. These disease-centred biobanks exist in France and used to be governed by academic or private institutions, mostly patient organizations such as Association Française contre les Myopathies, France Parkinson–France Alzheimer, and Biobanque de Picardie.[2]

French law does not refer to these institutions as 'biobanks', but nevertheless the legislator recently provided them with a relatively interconnected legal framework.[3] While specific technical rules relate to the activities of biobanks (under what conditions a biobank can be established, and under what conditions samples can be given to it), the general principles of 'French-style' bioethics, including the principles of 'non-property', informed consent, and data confidentiality, also apply to them. We will show that French legislators have grappled with and ultimately found solutions to the principal issues raised, thus paving the way for sustained biobank operations in France.

Despite this complex legal framework, the law has not yet covered all the questions arising from the biobanking activities, notably donor rights, such as ongoing monitoring of the use of biological material and related data donated for research purposes, researcher access rights, and current issues relating to ownership.

[1] See UK Biobank 2009; Estonian Gene Biobank 2009; deCode Genetics Company 2009; Swedish National Biobank Programme 2009.

[2] See Banque Genéthon de l'Association Française des Myopathies 2009; Banque de Tissus pour la Recherche de l'Association Française des Myopathies 2009; France Parkinson–France Alzheimer 2009; Biobanque de Picardie 2009.

[3] In Italy, rules for biobanking are based on different principles and dispositions (see Chapter 12), whereas in Switzerland, the aim is too enact an overarching human research law (see Chapter10).

Therefore we will explore the French legal framework for biobanks used in research, and try to shed light on a number of unresolved issues.

11.1 A continuously evolving legal framework

It can be stated without exaggeration that the legal framework in France has been continuously evolving ever since legislators became aware of the importance of biobanks. They started by issuing successive laws too numerous to describe here, then built a clearer framework with the Bioethics Act of 6 August 2004 (Thouvenin 2005a,b,c). This law echoed the Biological Resources Centres (BRC) promotion by a BRC committee established in 2001 by the Ministry of Research, under the aegis of OECD, in order to foster the biotechnology industry (Ministry of Research 2009).

Legislators attempted to address three core biobank issues to further their development.

1. Biobank sustainability depends on long-term funding because of high infrastructure and operating costs (Dagher 2006). This explains why the Public Health Code opened biobanking activities to the private sector (Article L1243-4 of the Public Health Code).
2. Broader access to biological materials is required because of the rapid development of molecular biology. Indeed, according to the Bioethics Act, not only biological material obtained from health-care facilities, but also surgical waste including placentas and fetal tissue, can be used for scientific purposes (Thouvenin 2005a,b,c). It is even possible to obtain exemptions, granted under very strict conditions, which allow research to be performed on frozen human embryos that are no longer needed for conception purposes.
3. Clearer provisions for all the stakeholders involved in biobanking are needed to build a secure legal basis, through the administrative monitoring of biobanking activities in addition to the well-known French bioethics principles providing donors with ethical and legal guarantees.

We will focus on the last of these provisions relating to biobanking activities since they constitute the core of biobanking. In concrete terms, there are two corpuses of rules: rules specific to biobank administrative status that can be found in the Public Health Code, and bioethics principles found in both the Civil Code and the Public Health Codes.

11.1.1 Specific rules for the administrative status of biobanks

The Public Health Code distinguishes between therapeutic and scientific biobanks, and we will focus on the latter as they are the subject matter of this chapter.

Articles L1243-3 and L1243-4 of the Public Health Code deal with biobanks created by researchers for their own research purposes, and biobanks created to store and supply other researchers with biological samples, respectively. The distinction is somewhat artificial since, in practice, these activities frequently overlap. These Articles clearly and strictly regulate the administrative status of a biobank from the moment that it is created until the moment that it stops functioning:

When a biobank is created by researchers for their own research, such as in the case of a research unit of the Institut National de le Recherche Médicale (Inserm) or the Institut National du Cancer (INCa), which manages a tumour bank network in France,[4] it must be declared to the Ministry of Research under the terms of Article L1243-3 (as well as to the regional hospitalization office if the biobank is located in a health-care facility). We will not delve into the intricate details of the

[4] See INCa 2009.

declaration process, but it is useful to highlight the important role played by the Comité de Protection des Personnes (CPP), which is tasked with protecting individuals in the biomedical research field. Indeed, in the event of an unfavourable CPP opinion, researchers can neither create a biobank nor declare it to the Ministry of Research. Finally, the Minister of Research wields administrative power, granting or refusing licences for the creation of biobanks.

When a biobank is created to store biological samples and supply them to external researchers, the head of the biobank has to apply to the Ministry of Research for authorization under the terms of Article L1243-4 (again, additional authorization is required from the regional hospitalization office if the biobank is located in a health-care facility). This type of biobank can be a public facility such as NeuroBioTec, a private non-profit facility such as the Banque de Tissus pour la Recherche or the Biobanque de Picardie, or a private company such as RNTECH.[5] It is important to note that authorization is required whether or not biobank transfers are free of charge.

11.1.2 **Bioethics principles specifically adapted to fit biobanking activities**

Biobank activities are based on the core principles of Bioethics Acts[6], which cover body part removal as well as biological sample flows. Do these principles hinder biobanking activities through their strict monitoring of the way that biological samples are removed and used, in line with ethical guarantees such as donor consent, data confidentiality, and the French-style 'no-property' principle? How can biobank activities be considered legal in light of these principles, particularly the 'no-property' principle which is currently the subject of heated debates in France and elsewhere? Indeed, biobanking activities do give rise to commercial activities—although they store human biological samples free of charge, they also transfer human biological samples for a fee and reap many of the benefits brought by research.

We will first clarify what the 'no-property' principle actually entails, and then focus on other principles related to donor safeguards that exist in most countries, such as the right to consent and right to privacy through data confidentiality. We shall see that they have been modified to fit biobanking activities.

The 'no-property principle'[7] is strongly emphasized in the Civil Code. It is incorporated three times and its provisions are 'public order' rules. Article 16-1, paragraph 3, of the Civil Code[8] states: 'The human body, its elements and its products may not form the subject of a patrimonial right'. This means that human body parts or tissues are excluded from the heritage[9] of the donor

---

[5] See RNTECH 2009.

[6] These principles are from the *Bioethics Act no. 94-653 relating to the respect of the human body*, and *Bioethics Act no. 94-654 relating to the donation and use of human body elements and products, to medically assisted procreation, and to prenatal diagnostics* of 29 July 1994 that have been modified by *Act no. 2004-800 relating to bioethics* of 6 August 2004.

[7] More exactly, *le principe de non-patrimonialité*, which means that which cannot be included in somebody's heritage.

[8] See also Article 16-5 of the Civil Code which provides for the public order cancellation of patrimonial agreements: 'Agreements that have the effect of bestowing a patrimonial value to the human body, its elements or products are void'; Article 16-6 of the Civil Code: 'No remuneration may be granted to a person who consents to an experimentation on himself, to the taking of elements from his body or to the collection of products thereof'.

[9] The heritage is a unit of rights and obligations of a person related to goods, which have an economic value. For example, the right to property, the right to succeed somebody, the duty for children to pay their dead father's debts is patrimonial.

or anyone else, such as the physician who collects them for scientific use. Thus human biological samples cannot be appropriated by anyone through transfer, be it free of charge or for sale; for example, the donor is not entitled to sell blood samples to either the physician who collects them or anyone else. However, by virtue of Article L1243-4 of the Public Health Code, as soon as the bodily material is collected and stored in a biobank for scientific use, it is possible to transfer it, i.e. to transfer the property right of the material. Thus the legislation implies that stored biological material is considered as a good which can be appropriated.

It is still unclear whether stored biological samples can be considered as goods since the legislator has not defined precisely when they are transformed into goods. This issue gives rise to very divergent viewpoints. Some consider samples to be appropriated goods as long as an innovation is derived from human biological material. The famous John Moore case[10] is representative of this first viewpoint. Others believe that samples can be considered as appropriated goods from the moment they are stored in biobanks because of the high degree of sophisticated technical handling they require. In this vein they can point to the decision by Australia's High Court of Justice in the case of *Doodeward* v. *Spence* in 1908,[11] which implicitly entitled a researcher to property rights over a fetus stored in formaldehyde by giving him the right to prosecute the person who had stolen it (see also Chapter 9, section 9.4.3). The judges found that the researcher had undertaken a real storage process (Skene 2002).

Although the case was decided a long time ago, it seems to fit current scientific storage work in scientific biobanks. Indeed, biological sample storage requires a high degree of technical handling, which can be expensive, as in the case of DNA and RNA extraction for scientific needs, or when samples are conditioned in the form of flakes or aliquot tubes. However, most biobank managers claim that they do not sell biological samples per se, but just logically charge for sample preparation, storage and transfer activities. Even if we posit that, in certain circumstances, be likened to the object of a transaction, it remains a 'human object' since it cannot be collected or used without prior donor consent (Bayer 2003).

The informed consent principle relates to the collection of human body elements and products, and is contained in Article L1211-2 of the Public Health Code. It is a sine qua non for the collection of human body samples for medical or scientific use. We would like to point out that this system of consent had to be adapted to biobanking activities, where the use of biological samples has changed over time because of secondary use, for two main reasons. First, biobanks store biological samples for very long periods during which research purposes different from those to which the donor originally consented may emerge. Secondly, the considerable progress in the field of molecular biology has meant that elements of the human body are now considered to be prime research material. Thus material obtained for medical purposes in the health-care sphere (such as diagnosis or medical follow-up) is now being recovered by researchers for scientific purposes. It is for these reasons that the Bioethics Act ensures that donors are informed of any secondary uses, and that they have a right to object to them. Both informed consent scenarios were written into the law: changes in research uses for samples (Article L1211-2 §2), or shifts from medical to scientific purposes (Article L1245-2).

Biological samples donated for scientific research would have considerably less value if they were not combined with donor health data. The legislator is aware of this and has provided a legal apparatus both to facilitate access to data and to guarantee its confidentiality. Indeed, such access

[10] *Moore* v. *Regents of the University of California*, 793 P. 2d 479 9 (Cal. 1990).

[11] *Doodeward* v. *Spence* [1908] 6 CLR 406.

could have led to breaches in professional secrecy, since physicians would be transmitting data to researchers and violating donor privacy.

The data confidentiality principle was set out in the Loi Informatique et Libertés (Act Relative to Data Protection and Liberties) of 6 January 1978 and modified on 6 August 2004. Briefly, this law lists three steps with which researchers must comply: applying for data access to the Commission Informatique et Libertés (CNIL),[12] an independent administrative authority protecting privacy and personal data, informing donors, who may object to their data being processed, and, finally, data coding.

1. Accessing data requires a declaration made to the CNIL for gathering data related to the samples collected in order to create the biobank collection (Article 23). In addition, an application must be submitted to the CNIL for authorization when the collected data are to be used for specific scientific research (Article 54).
2. Data access also requires that donors are informed of the potential research uses of their data and of their right to object without having to 'justify it on legitimate grounds' (Articles 54 and 57).
3. Data coding is, of course, required when the data are still linked to donors, making identification possible. However, the law remains unclear on the coding process, possibly because such details are provided on a case-by-case basis to the CNIL by researchers. Such uncertainties can be problematic, especially for data transfers in third countries where data derogation can occur (Article 55); nevertheless it is possible to resort to European Commission standard contractual clauses (see Chapter 13, section 13.7.3).

11.2 Major (outstanding) legal issues

Despite the existence of an elaborate legal framework, major legal uncertainties remain, such as those relating to (1) researchers' biobank access and property rights, and (2) the user rights of donors with respect to their biological samples and related data.

11.2.1 Access to biobanks and ownership issues

Biobank access and ownership issues are vital not only to researchers and to facilities housing biobanks, but also to actors involved in for-profit research. We will first examine access to biobanks, and then biobanking ownership issues:

1. Access to biobanks is crucial to researchers, whose work depends on conditions of access (free of charge or not, delayed or not) in a highly competitive scientific environment. French legislators did not foresee the rapid expansion of biobanks and thus did not provide any specific legislative solutions to these issues. As far as we know, professionals are not penalized, since they can refer to (a) professional guidelines which facilitate access to biobanks and (b) to contractual agreement models which formalize and organize access to biological sample collections.
    a. Professional guidelines set out recommendations for stakeholders through procedures that strike a balance between access to biobanks and their long-term sustainability. Biobanks are a source of knowledge; thus it would be highly detrimental to scatter these collections with chaotic researcher access protocols. Interestingly, these guidelines provide for different models, including completely closed access, as in the case of a biobank

[12] Commission Informatique et Libertés 2009.

managed by GlaxoSmithKline, a company which conducts pharmacogenomic research on collections obtained from clinical trials, and both free and fee-charging open access, such as the protocols used by non-profit private facilities (e.g Biobanque de Picardie, Banque de Tissus pour la Recherche, and Banque Généthon) with free access for academic researchers who created the collection and fee-paying access for pharmaceutical companies through specific agreements. Clearly, stakeholders are developing new approaches to biobank access; another example is the emerging trend of granting 'temporary priority rights' to gain access to biobanks, as currently bestowed by the Association Française contre les Myopathies, Biobanque de Picardie, and the Institut National de la Recherche contre le Cancer.

b. Lastly, we can focus on contractual agreement practices explicitly required by the Public Health Code for biological sample transfers, but for which no pattern has been provided for professionals who require access to human biological material. Most national research institutes have created their own contract models (Bellivier and Noiville 2006). We can distinguish between two types of agreement: material transfer agreements (MTAs) and scientific collaboration agreements. The distinction essentially relates to the subject matter of the contract. In MTAs, it consists of supplying biological material in a safe and transparent way to facilitate sample flows, whereas in scientific collaboration contracts the supply is merely a part of the principal contract objectives, which aim above all to lay the groundwork for scientific collaboration, including rights and obligations for the contracting parties, notably benefit-sharing in case the research leads to financial gain. We should also note that MTAs are often more standardized than collaboration contracts, since *intuitu personae* logically have greater influence in drawing up collaboration contracts than MTAs. Moreover, MTA standards are so useful to researchers that most reputable research institutions (*Assistance Publique des Hôpitaux de Paris (AP-HP), Institut National de la Recherche Médicale (Inserm)*, etc.), which manage their own biological standards, have adopted greater standardization to the point where their MTA standards can be compared with those of foreign research institutions.

2. Questions about biobanking ownership in France remain because the law is largely silent on the matter. It is unclear whether biobankers or researchers are entitled to be owners of human biological samples (for a distinction between different legal paradigms of human tissues, see Chapter 9). One thing that is certain is that, by virtue of Article 16 of the Civil Code, donors are not the owners. However, most observers agree on the limits of property rights, given that they are exclusive and thus inherently unfair.

Unlike recent US case law, in which the judge in the *Catalona* case[13] gave the property rights over a collection of tumour samples to the academy holding them, and not the physician who had

[13] The *Catalona* case deals with a conflict of interest between Dr Catalona and Washington University, which housed the biological sample collections built up thanks to the consent of Dr Catalona's patients suffering from prostate cancer. Research had produced promising therapeutic products to help cure prostate cancer. Dr Catalona decided to leave Washington University, where he had worked and stored his collection for years, but the university argued that it had property rights over the collection since it had managed and looked after it during that time. The judge ruled that, according to Missouri state law: Dr Catalona could not be the owner because his employment contract stated that he had given up such rights; according to the *Moore* case, donors could not be the owners (Washington University in St Louis 2009). The judge's ruling is obviously the least worst solution because recognizing physicians' property rights over collections would tend to lead to their dispersal over time. Nevertheless, it is unfair to

collected them (for a critical appraisal, see Chapter 9, section 9.4.2), most influential French public policy authorities such as the National Ethics Committee have objected to biobank ownership. In opinion number 77, the National Ethics Committee recommended that there is no ownership and that biobank managers should be considered 'guardians' because storage cannot be likened to acquisition or ownership (CCNE 2009). Despite these intense ongoing ethical deliberations among biobanking policy-makers and stakeholders, substantive law on the matter remains unclear.

It is legitimate for donors to be informed of the uses of their biological samples and of the related data because research can rapidly lead to different scientific assumptions and approaches (see Chapter 2). However, we shall see that questions remain on donor information and consent related to (1) secondary use and (2) research results.

1. As illustrated above, the principle of consent has been adapted to biobanking practices by providing donors with a kind of uninterrupted consent relating to the use of their biological samples: donors are informed of potential secondary uses for their samples and related data, and of the right to object to them. Nevertheless, a closer examination of Article L1211-2, paragraph 2, of the Public Health Code[14] removes much of the meaning from informed consent. Implementation of this article suffers from at least three shortcomings: First, it can be disregarded when a donor cannot be found or he/she is deceased, which is a frequent occurrence in research which can last for years. Secondly, when the Ethics Committee (Comité de Protection des Personnes) is asked to render an opinion by the person in charge of research, it can release the latter party from the obligation to inform donors when 'it considers the information to be unnecessary'. We can infer from the wording that the Ethics Committee has discretionary power and does not need to justify such exemptions. This results in preferential treatment for those researchers who are allowed to bypass donor notification. The legal situation is all the more worrying given that French public research organizations, in a hurry to start their research as quickly as possible in an intensively competitive scientific environment, skirt informed consent requirements by falling back on broad consent forms, enabling them to use biological samples for purposes that are different from the initial ones without informing the donors. Thirdly, it would be judicious to consider creating a notification period so that the right to object can be properly exercised. For this reason the Institut National du Cancer and Assistance Publique des Hôpitaux de Paris have issued recommendations for clinicians and research organizations which emphasize donor notification and confirming donor consent to non-diagnosis or non-health-care uses: 'it is necessary to ensure that patients be provided with a sufficiently long notification period to think about, question and express their concerns, and possibly their objections…' (INCa 2009).
2. What can we say about donor notification regarding benefit-sharing and research results? It is widely known that these issues have become the cornerstone of biobanking activities (see Chapter 5, section 5.3). However, the question of the legitimacy of donor rights claims to the innovations resulting from research, even though academics and industry could not have developed benefits without them, still remains open. On the one hand, donors merely provide

physicians who spend years building up collections. Moreover, there is nothing to stop a university from breaking up or even destroying such a collection.

[14] Art. L1211-2, para. 2: '… It can be dispensed from the obligation to inform when finding the individual in question is impossible, or when one of the advisory committees in charge of the protection of individuals…, in consultation with the person responsible for the research, does not believe the information to be necessary' (own translation).

an unfinished resource with low inherent value and a potential high added value—value is added because of, and in proportion to intellectual, technical, and industrial investment. On the other hand, removal of body elements is undertaken by the national public health system and funded by the community, which share a legitimate claim to the resulting benefits (for a discussion of this issue drawing on the concept of solidarity, see Chapter 6, section 6.3).

Interestingly, donors in both France and other countries (Boris 2007) do not appear to pay close attention to these issues (Caze de Montgolfier 2007). They are far more concerned with being kept informed of the uses of their biological samples and related data, and that their samples and data are not misused for commercial or unethical purposes. Indeed, most donors are patients suffering from a serious illness who have placed high hopes in research. In light of this, the French legislator has left out a critical tool in the field of notification rights—a system which shares research results and takes donor participation into account. Currently, information about 'overall results' is only provided to individuals involved in clinical trial research, and not to the donors of biological samples. Moreover, results are only shared at the end of the research period. Setting up a continuous flow of information in biological sample research would be a way of keeping donors informed and showing that their involvement is highly valued. However, it is difficult to provide adequate individual donor information insofar as the research is essentially statistical, but most international research programmes have sufficient funding to enable the public, especially research participants and their families, to follow the results through regularly updated websites.

11.3 **Conclusion**

We have seen that biobanking activities give rise to issues that French law has not yet ruled on, such as researchers' rights relating to the use and the added of biological collections, as well as private donor rights to consent and privacy.

Beyond private interests, biobanking practices give rise to collective interests among the donor community. Indeed, the more such practices occur, the more donors are empowered to become 'biological citizens' (Mayrhofer 2008) because large-scale biobanking activities are increasingly taking place in national and/or transnational networks. The collective involvement is a dimension of public health policy which should be taken into account: biobanking activities, which on the one hand have a significant impact on national health-care systems by generating epidemiological data to prevent diseases, are potentially in conflict with donor privacy; on the other hand, access to therapeutic developments by donors and society as a whole is also be an issue of major importance.

While these issues have been taken into account on ethical grounds in international research projects, they have not yet been taken into account on legal grounds at the national and international level. The only legal instrument that we are aware of which addresses this concern is Recommendation Rec (2006) 4 in *Research on biological materials of human origin.*[15]

References

Banque de Tissus pour la recherche de l'Association France Myopathie. (2009). Available at *http://www.institutmyologie.org/ewb_pages/r/recherche_banquetissus_equipe.php* (accessed 1 April 2009).

[15] Article 19, Oversight of population biobanks, section 1: 'Each population biobank should be subject to independent oversight, in particular to safeguard the interests and rights of the people concerned by the biobank activities'.

Banque Généthon de l'Association France Myopathie. (2009). *Presentation.* Available at http://www.genethon.fr/index.php?id=82 (accessed 1 April 2009).

Bayer, E. (2003). *Les choses humaines.* Law Thesis, Université de Droit de Toulouse.

Bellivier, F. and Noiville C. (2006). *Contrats et vivant.* Paris: LGDJ.

Biobanque de Picardie. (2009). *Presentation.* Available at http://www.biobanque-picardie.com/ (accessed 3 April 2009).

Boris, B. (2007). Tissue banking: patients' views and expectations in a regulated environment—does this help the patient? Part 2: Patient views and expectations. *Biopathology*, **74**, 223–6.

Caze de Mongolfier, S. (2007). *Collecte, stockage, et utilisation des produits du corps humain dans le cadre des recherches en génétique : état des lieux historique, éthique et juridique, analyse des pratiques au sein des biothèques.* Available at: http://www.ethique.inserm.fr/inserm/ethique.nsf/1d74e2daceb53478c1257061005 64aea/933a7f5b352f398ac12570a500515255/$FILE/Pr%C3%A9sentation%20orale.pdf (accessed 23 May 2007).

CNIL (Commission Nationale Informatique et Libertés). (2009). *Loi Informatique et Libertés (of January 6, 1978 amended by Act of August 6, 2004 relating to the protection of personal data).* Available at http://www.cnil.fr/fileadmin/documents/uk/78-17VA.pdf (accessed 25 May 2009).

Dagher, G. (2006). Interview with Georges Dagher, Biological Collection Manager of French National Institute of Medical Research. *Journal Libération*, 25 November 2006,Available at http://www.liberation.fr/transversales/weekend/219254.FR.php (accessed 1 April 2009).

deCODE Genetics Company. (2009). Available at http://www.decode.com/(accessed 1 April 2009).

Estonian Gene Biobank. (2009). *Presentation.* Available at: http://www.geenivaramu.ee/(accessed 3 April 2009).

Fletcher, A. (2008). Governing DNA. Prospects and problems in the proposed large-scale United States population cohort study. In H. Gottweis and A. Petersen (eds), *Biobanks: governance in comparative perspective*, pp. 114–15. London: Routledge.

France Parkinson–France Alzheimer. (2009). *Neuro-CEB, a brain tissue bank in the field of neurological diseases.* Available at http://www.arsep.org/?mod=cms&ID=140&lang=en (accessed 1 April 2009).

Gottweis, H. and Petersen, A. (2008). Introduction. In H. Gottweis and A. Petersen (eds), *Biobanks: governance in comparative perspective*, pp. 5–6. London: Routledge.

INCa (National Cancer Institute). (2009). *Les tumorothèques hospitalières, recommandation à l'usage des cliniciens et des chercheurs.* Available at: http://www.e-cancer.fr/Espace-tumorotheques/Introduction/op_1-it_681-la_1-ve_1.html (accessed 24 May 2009).

Mayrhofer, M. (2008). Patients' organization as (un)usual suspects in biobanking. In H. Gotweiss and A. Petersen (eds), *Biobanks: governance in comparative perspective*, pp. 71–87, London: Routledge.

Ministry of Research. (2009). *About biological resource centres.* Available at: http://www.Recherche.gouv.fr/discours/2001/crbiod.htm (accessed 1 April 2009).

RNTECH Company. (2009). *Presentation.* Available at http://www.rntech.com (accessed 19 April 2009).

Skene, L. (2002). Ownership of human tissue and the law. *Human Genetics*, **3**, 145–8.

Swedish National Biobank Program. (2009). Available at http://www.biobanks.se/(accessed 1 April 2009).

Thouvenin, D. (2005a). La loi relative à la bioéthique ou comment accroître l'accès aux éléments biologiques d'origine humaine. *Recueil Dalloz*, **2**, 116–21.

Thouvenin, D. (2005b). La loi relative à la bioéthique ou comment accroître l'accès aux éléments biologiques d'origine humaine. *Recueil Dalloz*, **3**, 172–9.

Thouvenin, D. (2005c). Les banques de tissus et d'organes: les mots pour les dire, les règles pour les organiser. *Les Petites Affiches*, **18**, 31–47.

UK Biobank. (2009). *Welcome to UK Biobank.* Available at: http://www.ukbiobank.ac.uk/(accessed 3 April 2009).

Washington University in Saint Louis. (2009). *Medical Public Affairs: statement on U.S. Supreme Court's Denial of Certiorari in case involving ownership of tissues donated for research.* Available at: http://prostatecure.wustl.edu/(accessed 4 May 2009).

Chapter 12

Biobanks: ethical and legal aspects of the collection and storage of human biological material in Italy

Antonio G. Spagnolo, Viviana Daloiso, and Paola Parente

12.1 Introduction

Collecting biological samples has always been an important tool for research into human diseases, for improving understanding of patterns of illness, and for searching for new pharmaceutical compounds and novel therapies. The rapid pace of change in medical research, in particular in the field of genetics, in recent decades has produced anxiety about an increasing risk of invasion of privacy and genetic discrimination. Although the public strongly support scientific breakthroughs, promising better medical diagnoses and treatments, it is imperative to guarantee the privacy and confidentiality of sample donors.

The aim of this chapter is to outline the Italian ethical and legal position that has emerged in a number of important official documents. In particular we will deal with the following topics:

- the definition of a 'biobank' within national documents
- the content of donor information and the form used to obtain consent
- modalities adopted to safeguard privacy and confidentiality about data and results
- ownership of samples.

12.2 Overview of Italian approach to biobanks

Currently, an enormous number of human biological samples and data have been collected in Italy by various research groups with different levels of safety and computerization. In such a situation, there is a potentially high risk of losing these data and endangering the sample donors. In Italy, there is a lack of specific rules and regulations concerning biobanks. However, several committees have studied the matter and three documents have been produced:

- *Guidelines for genetic biobanks* issued by the Italian Society of Human Genetics (SIGU) and the Telethon Foundation (Italian Society of Human Genetics 2004);
- *Guidelines for the establishment and accreditation of biobanks* issued by the National Committee for Biosafety and Biotechnology (Ncbb), an advisory body of the Office of the Prime Minister (Italian National Committee for Biosafety and Biotechnology 2006);
- *Biobanks and research on human biological material* issued by the National Bioethics Committee (Nbc) (Italian National Bioethics Committee 2006).

The first and the second of these documents are guidelines aimed to improving the definition and evaluation of specific applications concerning biobanks; the third is a survey of the ethical and legal implications of the storage of human samples. In particular, the third document was drafted by comparing the *Guidelines for the establishment and accreditation of biobanks* (Italian National Committee for Biosafety and Biotechnology 2006), with the paper *A draft recommendation on research on human biological material* issued by the Steering Committee of Bioethics (CDBI), (Steering Committee of Bioethics 2005). The Italian scientific community considers these documents as an important point of reference for those involved in research.

12.3 **The definition of biobanks and stored samples**

The SIGU–Telethon Foundation and NCBB guidelines first define the meaning of the term 'biobank' and establish what kinds of sample can be stored in a biobank. The working group set up by SIGU and the Telethon Foundation have proposed guidelines for the creation, maintenance, and use of genetic biobanks (Italian Society of Human Genetics 2004). This document is the basis of further studies and discussions aimed at producing national guidelines, or at least binding rules for genetic biobanks, as soon as possible. In this document, a biobank is defined as a specific institution characterized by the collection, storage, and exchange of samples from individuals and families with genetic pathologies, population groups with a high frequency of persons who are carriers of or are affected by genetic diseases, and populations with genetic characteristics identifiable through susceptibility genes.

The samples stored in a biobank can be human cell lines, DNA samples, and transgenic or engineered material. All biobanks must be organized to ensure the privacy of the donor, the quality of the sample, appropriate storage of the sample for as long as possible, and the correct use and distribution of the sample.

The need to improve national research and to adapt it to European standards prompted the government to ask the Ncbb to produce an opinion on biobanks. In April 2006, the NCBB issued a document entitled *Guidelines for the establishment and accreditation of biobanks*, with an appendix entitled *Guidelines for biobank certification* (Italian National Committee for Biosafety and Biotechnology 2006). These guidelines are based on European and international documents such as the regulations adopted for the European biobanks at the University of Maastricht, where the term 'biobank' refers to 'a service unit for reliable storage and automated management of biological material and corresponding data, complying with the Code of Good Use and the Code of Good Behaviour, and any additional guidelines of the Medical Ethics Committee and the University' (Stelma 2003). Another document within the European context is the European Council Recommendation on human tissue banks, which defines such a bank as 'a non-profit organization' that must be officially recognized by the competent health authority of the member state, and must guarantee the treatment, preservation, and distribution of the material (Council of Europe 1994). Following these documents, the NCBB defines a 'biobank' as a 'non-profit service unit'[1] created for the collection and storage of human biological material used for diagnosis, the study of biodiversity, and research. The specificity of biobanks requires that it must be possible to link the stored samples to the personal, genealogical, and clinical data of the subject from whom the deposited material has been collected (Italian National Committee for Biosafety and Biotechnology 2006). These guidelines represent a benchmark for more homogenous national legislation in this field, and also demonstrate a new approach to solidarity between groups and generations based on the voluntary sharing of samples and information as a joint resource to

[1] This definition is similar to the French approach (see Chapter 11).

benefit society.[2] The NCBB also specifies the type of samples that can be stored: samples can be derived from all parts of the body including surgical residues; only hair, nails, placenta, and body waste products are not considered to be human tissues. Moreover, the guidelines point out that biobanks should not preserve biological material that is already regulated, i.e. organs for transplantation, blood and blood products for transfusion purposes, embryos, and sperm and oocytes for assisted reproduction.

Following the Recommendation of the Council of Europe (Council of Europe 2006), the NBC issued a document, approved at the plenary session on 9 June 2006, entitled *Biobanks and research on human biological material* (Italian National Bioethics Committee 2006). This document deals with crucially important issues concerning the bioethical implications of sampling and storage of human tissue, guaranteeing human dignity and rights and, at the same time, taking the requirements of research into consideration. Furthermore, the NBC tried to identify ways of safeguarding the interests of everybody involved in the research, as well as the interests of society, to facilitate the search for better scientific knowledge and to improve the common good, entailing well-being and health, through research. In particular, the first ethical and scientific step before undergoing any research protocol should be the assessment of the risks and benefits to the donors and the need to guarantee their anonymity and privacy.

The term 'biobank' appears in only two Italian national legal documents: law no. 40/2004—*Norms on medically assisted fertilization* (Italian Parliament 2004), and the decree *Urgent measures on umbilical cord stem cells* (Ministry of Health 2007). According to the former, cryoconservation of embryos is not allowed. Before the law came into effect a specific decree was issued covering all those embryos which are cryoconserved but discarded or which are not designated for transfer to the womb in fertility centres. It has been established that discarded embryos should be stored in a public national biobank located in the Ospedale Maggiore in Milan. The decree requires that (1) cord blood for solidary allogenic purposes should be stored in a certified biobank (currently there is a network of about 20 biobanks), (2) storage is not allowed in private biobanks for dedicated autologous purpose, and (3) autologous storage and use is allowed only in specific situations.

12.4 Informed consent and information form

Currently, there is no comprehensive and unique legislation for informed consent in Italy, and rules are derived from different principles and dispositions (for the value of informed consent in the context of biobanking, see Chapter 2). The first principles to refer to are based in the Constitutional Charter; Article 13 states the inviolability of personal freedom, and Article 32 affirms the need of personal informed consent before any medical intervention. The Italian Constitutional Court has recognized that the latter embodies the freedom of everyone to dispose of their own body.[3]

More recently, the need for consent has also been called for in article 5 of the Convention on Human Rights and Biomedicine (signed by the Italian Government in Oviedo in 1997, and ratified in Italy on 28 March 2001) (Council of Europe 1997). There are other dispositions within the national legislation, but by and large they remain general. Other more general rules can be derived from the Italian Code of Medical Deontology with respect to information and consent (Italian National Federation of the Medical Order 2006). Article 33 of the Code states that the physician

---

[2] For a discussion of the meaning of the term 'solidarity' in the context of biomedical and genetic research, see Chapter 6, section 6.3.

[3] Italian Constitutional Court 1990, Sentence n. 471/1990.

is required to inform the patient, using clear and understandable language; article 35 states that the physician must not undertake any diagnostic or therapeutic activity without the express well-informed consent of the patient.

Although there is no specific law, it is possible to identify some key elements that must be strictly observed to obtain true informed consent: the way that the information is passed to the subject; who can consent and what to do when someone is unable to give consent because of mental, physical, or legal incapacity; freedom of choice and the possibility of withdrawing consent. *Guidelines on biobanks* tries to answer to these important questions.

The proposed guidelines for genetic biobanks (Italian Society of Human Genetics 2004) state that informed consent for the collection of samples must also include consent for storage of samples and their possible use for further research and/or diagnosis. This should be discussed with donors before samples are taken. At this time, researchers should give clear and complete information in order to obtain 'true' informed consent (the difficulties of obeying this rule are addressed in Chapter 2). If the donor has any doubts, he/she should be allowed to consult third parties not involved in the research project (ibid.:section 3.1). When the collection of samples involves an entire population, consent should be requested from each participant and not simply from a representative of the population; this is significantly different from the procedure with the deCODE project in Iceland. Clear and true information is not only an essential requirement defined in all guidelines, but should also be guaranteed by an ethics committee for every single research project, to ensure the right of any individual to refuse participation in a study. In the form used to obtain the consent, all the elements regarding the management of the stored biological samples should be presented; it should be clear that the sample can be used for further studies with research and/or diagnosis purposes concerning the same pathology (if the research has a different aim, further specific consent should be obtained from the subject); the kind of information that could be obtained by using the sample should be well defined; and the potential personal or societal benefits should be specified (for an overview of these different interests, see Chapter 5). As far as privacy is concerned, the sample should be collected and stored in such a way as to ensure data confidentiality. All the procedures used in the management of the collected data should be clarified bearing this in mind. Moreover, the donor's right to withdraw consent should be clearly stated in every case. If a donor withdraws consent, all the collected information pertaining to that sample must be destroyed (ibid.:section 3.1.1). In contrast, the guidelines of the Swiss Academy of Medical Sciences (SAMS), for example, do not require destruction of results derived from samples (see Chapter 10, section 10.3.2).

There are further potential sources of biological samples which represent 'particular situations': samples obtained from individuals who are unable to provide consent, samples for which informed consent is not possible, and samples obtained from abortions. In the first case (samples collected from minors or those lacking the capacity to consent because of mental illness), according to the Oviedo Convention, Italian legislation[4] allows consent to be obtained from a legal representative. When it is not possible to obtain any form of consent, use of the sample is acceptable only if the potential benefits for the individual and society are very significant and satisfy all the measures established for samples with consent. With regard to samples collected from aborted fetuses, the guidelines make a distinction between spontaneous abortions and deliberate abortions. In the case of deliberate abortions, it is important to ascertain whether the abortion was performed after genetic diagnosis of pathologies. With regard to all these types of abortion, whether there is a request for further investigation aimed at identifying the fetal pathology causing

[4] Italian Legislative Decree n.135/1999, Article 2.

the termination of the pregnancy, or if they are required for research, the biological samples should be stored according to the NBC recommendations. In any case, these possibilities are only acceptable when consent has been given by the parents and the independence of those who terminated the pregnancy (the health personnel) and those performing the experiments has been verified. A request for the mother's consent can only be omitted when the biological material is obtained from an abortion performed during the first 12 weeks of the pregnancy and the sample is anonymized.

12.5 **Privacy and confidentiality regarding data and results**

A problematic aspect of biobanks is the link between samples and data. Identification of samples and data is a key problem both because researchers need to share data and, at the same time, donor privacy must be ensured.

A donor's right to privacy is defined in Article 21 of the Charter of the Fundamental Rights of The European Union (European Union 2000), Article 21 of the Oviedo Convention (Italian National Federation of the Medical Order 2006), and the recommendations of the European Group on Ethics on genetic testing that 'all medical data, including genetic data, must be afforded equally high standards of quality and confidentiality at all times' (European Commission 2004). The scientific community has made many proposals employing different terminology to ensure confidentiality (Knoppers and Saginur 2005).

In Italy, confidentiality is regulated by the Personal Data Protection Code,[5] which defines the need to implement certain procedures with respect to access to and the communication of personal data, the possible involvement of family members, and the measures adopted for protecting confidentiality. Based on proposals by the European Commission (2004), samples can be classified into the following storage categories: identified; identifiable—further divided into single and double coded, and anonymized and anonymous. All identified or identifiable data, and also 'likely' identifiable data, are governed by the the Personal Data Protection Code. The need to guarantee the privacy and confidentiality of data, in particular in the field of genetics, serves the additional purpose of avoiding the risk of discrimination based, for example, on genetic characteristics.

Despite this, we must recall the double risk of anonymity, i.e. the 'irreversible stripping of identifiers from samples' (Knoppers and Saginur 2005), where the scientific value of the samples would be greatly reduced over time, and a possible benefit resulting from the research could not be assigned to the sample donor.

12.6 **Ownership of samples**

The debate regarding ownership of biological samples is still ongoing (see Chapter 9). The nub of the question can be summarized as follows. Whoever consents to the use of his/her sample gives the community a chance to promote health, for example by assisting research to understand and cure pathologies. How do we consider the act of donating a sample? Do we see it simply as a gift? As a rule, a gift cannot be taken back. If researchers using the sample find a therapy, who will benefit from the returned goods? What if the sample is subjected to commercialization? These questions are still awaiting an answer. in any case it is considered that the donated sample should not be a source of economic interest; on the contrary: it should guarantee the return of goods to

[5] Italian Legislative Decree n.196/2003. Available at: http://www.garanteprivacy.it/garante/document?ID=727068 (accessed 22 July 2008).

the society. Implementation of European Directive 2004/23/EC (European Parliament and the Council of Europe 2004) in the Italian legal system clarifies the issue: although the human body must not be subjected to economic exploitation, the economic benefits associated with the intellectual property deriving from the original work are recognized. Therefore the donor should be informed that the sample itself will never be used for profit, but that there is the possibility of economic advantage following indirectly from the use of the sample through the development of therapies or diagnostic tests which are in the interest of society.

12.7 **Conclusion**

It is well known that biobanks are not simply a 'case' for biological samples; rather, they are a resource showing an incredible potential to reveal vast amounts of information regarding the health of individuals and the risk of their developing disease and disability in the future. These issues involve the human as a sample donor and a subject of research. Thus a careful evaluation about the ethical issues involved cannot be avoided; on the contrary, this evaluation should be the first step—the prime mover—of every consideration involving biobanks.

Viewed in this light, two ethical implications are particularly urgent: (1) the ethical justification for informed consent; (2) the link between samples and personal data with its implications for the privacy of the donor or patient and for data confidentiality. Biobanks should protect personal data to avoid the risk of discrimination because of genetic characteristics by employers or insurance companies. Biobanks should represent not only a collection of 'human data', but a way of promoting health and giving more benefits to society. This can only be done when respect for the human person is recognized at all times.[6] Scientific research, when it comes to the sampling and storage of biological samples, must be based on respect for human values.

References

Council of Europe (1994). *Recommendation R(94)1 of the Committee of Ministers to member states on human tissue banks*. Council of Europe.

Council of Europe (1997). *Convention for the protection of human rights and dignity of the human being with regard to the application of biology and medicine: Convention on Human Rights and Biomedicine*. Oviedo.

Council of Europe (2006). *Recommendation Rec(2006)4 of the Committee of Ministers to member states on research on biological materials of human origin*. Available at: https://wcd.coe.int/ViewDoc.jsp?id=977859&Site=COE&BackColorInternet=DBDCF2&BackColorIntranet=FDC864&BackColorLogged=FDC864 (accessed 05 January 2010).

European Commission (2004). *25 Recommendations on the ethical, legal and social implications of the genetic testing*. Available at: http://ec.europa.eu/research/conferences/2004/genetic/recommendations_en.htm (accessed 3 January 2010).

European Parliament and the Council of Europe (2004). *Directive 2004/23/EC of the European Parliament and of the Council of 31 March 2004 on setting standards of quality and safety for the donation, procurement, testing, processing, preservation, storage and distribution of human tissues and cells, Official Journal L 102, 07/04/2004 P. 0048 - 0058*. Available at: http://eur-lex.europa.eu/LexUriServ/LexUriServ.do?uri=CELEX:32004L0023:EN:HTML (accessed 7 September 2009).

European Union (2000). *The charter of fundamental rights of the European Union (2000/C 364/01)*.

Italian National Bioethics Committee (NBC) (2006). *Opinion on biobanks and research on the human biological material*. Available at: http://www.palazzochigi.it/bioetica/pareri.html (accessed 05 January 2010).

[6] The implications of this concept are discussed in Chapters 2 and 3.

Italian National Committee for Biosafety and Biotechnology (NCBB) (2006). *Guideline for the establishment and accreditation of biobanks. Guideline for biobank certification.* Available at: http://www.palazzochigi.it/biotecnologie/documenti/7_biobanche_1.pdf (accessed 29 August 2009).

Italian National Federation of the Medical Order (FNOMCeO) (2006). *Italian code of medical deontology.* Available at: http://portale.fnomceo.it/Jcmsfnomceo/cmsfile/attach_3819.pdf (accessed 14 October 2008).

Italian Parliament (2004). *Norms on medically assisted fertilization, law n. 40 (19 February 2004).* Available at http://www.camera.it/parlam/leggi/04040L.htm (accessed 22 July 2008).

Italian Society of Human Genetics (SIGU) (2004). *Guidelines for genetic biobanks.* Available at: http://dppm.gaslini.org/documenti/GUIDELINES.pdf (accessed 13 January 2010).

Knoppers, B.M. and Saginur, M. (2005). The Babel of genetic data terminology. *Nature Biotechnology*, **23**, 925–7.

Ministry of Health (2007). *Urgent measures on umbilical cord stem cells.* Available at: http://www.ministerosalute.it/ministero/sezDettaglioDiario.jsp?id=44&anno=2007 (accessed 22 July 2008).

Steering Committee of Bioethics (CDBI) (2005). *A draft recommendation on research on human biological material.* Available at: https://wcd.coe.int/ViewDoc.jsp?id=961137 (accessed 3 January 2010).

Stelma, F.F. (2003). *Regulations European biobanks (rep. 04/2003).* Maastricht: Maastricht University Press.

Chapter 13

How to achieve 'free movement of tissue' in the EU research area[1]

Jasper Bovenberg

13.1 Prologue

In one sense, Antoni van Leeuwenhoek (1632–1723) can be considered the founding father of cross-border tissue banking. The Dutch autodidact scientist not only invented and manufactured his own microscopes, but also collected and analysed, among many other specimen, his own tissue, including the plaque on his teeth, his own blood, and his own sperm. He annotated his observations in great detail, arranged for their proper imaging, and shipped the results from Delft, The Netherlands, to the Royal Society in London. There his work was discussed, replicated, and subsequently published in the Society's *Philosophical Transactions*. Some of the samples he sent were so perfectly preserved that they can still be analysed today, almost 400 years after their submission to the Society's collection.

13.2 Introduction

At the time, van Leeuwenhoek's 'data and sample sharing' with the international scientific community did not raise any concerns or anxieties and did not met encounter any regulatory hurdles. Today, the widespread availability of annotated samples to the international scientific community seems just as pivotal to scientific progress as the availability of Van Leeuwenhoek's groundbreaking contributions. Yet while scientific research on human tissue is international by definition and human tissue samples have been shipped across European borders for centuries, these shipments are now a cause for anxiety and restrained by uncertainty over their legality. After a brief discussion of the root causes of this concern, this chapter will explore practical and legal pathways to overcome these concerns and uncertainty. In the absence of pertinent European Union regulation of the use of human tissue for research, I will draw an analogy between the international transfer of human tissue and the international transmission of personal data. This analogy suggests that cross-border shipments are permitted, provided that adequate safeguards are in place to address stakeholder concerns. Notably, analysis of the EU framework for cross-border transfers of personal data reveals a pragmatic public–private law approach to providing donors with adequate safeguards for cross-border shipments of their tissue in the form of an EU-approved set of standard contractual clauses. The analogous use of these clauses not only safeguards the interests of the institutions that are the parties to such contracts, but also provides enforceable civil remedies which empower individual donors whose tissue is being transferred to

---

[1] This chapter is based on the author's presentation at the Tiss.EU First Project Conference on 28 June 2008 in Göttingen, Germany.

secure their own interests, and thus allow free movement of human tissue for research in the EU research area.

13.3 Anxiety over tissue flows results from concerns over abuse

The root cause of the current anxiety and uncertainty surrounding the flow of human tissue across national (European) borders can be traced back to the interests of the stakeholders in cross-border sample flows. The stakeholders and their individual concerns can be roughly divided into three groups (Table 13.1).

The concerns of each stakeholder can be summarized as follows. A major concern for donors relates to the perceived potential of their tissue being cloned by a foreign scientist. Hence they do not want their sample to be shipped to, say, South Korea (Hwang *et al.* 2005). A second concern for many donors is the potential for their tissue to be used for commercial purposes by a foreign company (see Chapter 5, section 5.3 in this volume). As observed in a report by the Australian Law Reform Commission:

> Another thing that clearly emerged at the public forums is the atavistic or primal fear among members of the community about their genetic material being sent 'overseas' (again, often expressed as being 'sent to the US').... So, at almost every event, someone in the audience expressed concern about volunteering for an experiment at an Australian university research lab or teaching hospital, then finding that the research group had 'spun off' into a private biotech company, which then merged with or was taken over by American interests—and 'the next thing you know, your DNA is overseas'! (Weisbrot 2004)

Briefly, most donors do not want to find their freely donated DNA 'listed' overseas on the New York Stock Exchange. Also, donors have expressed serious concerns about the potential for sensitive data to be obtained from their sample and related concerns over the safe storage of their samples. As Nils Hoppe observes, 'it is inherent in the research methodology of many biobanks that a linkage to a specific individual is required for the research to be conducted and that the identifiability of the data subject hinges on the density of other data and contemporary technological means'.[2]

With regard to institutional concerns, institutions storing collections of samples must primarily honour any commitments, contractual and statutory, to the donors of these samples. This may force them to retain ownership or intellectual property rights of their collections and to provide adequate safeguards to protect data confidentiality. Also, as these samples are a finite resource, the institutions must take care to preserve the collection and set priorities for their limited use. With regard to national concerns, the investments that some nations have made in collecting

Table 13.1 Stakeholders and individual concerns

Stakeholder	Concerns
Donors	Private: cloning, commerce, confidentiality
Institutions	Institutional: depletion, investment, acknowledgment
Member states	National: biosafety, biopiracy, bioterrorism

[2] For an in-depth discussion of the types of risk that may occur in the context of biobanks and the (im)possibility of a proper balancing of these risks versus benefits, see Chapter 4.

samples lead them to want to use the samples as a unique selling point to attract business and scientists. To promote use within their borders, they prohibit export. National concerns over cross-border sample flows also relate to issues of epidemics and biosafety. The posibility that their citizens' samples may be exploited by foreign researchers (biopiracy) is another cause for concern. A final concern relates to the potential of tissue being used in acts of bioterrorism, as illustrated by the following event. In 2007, a report submitted to Vladimir Putin, then president of Russia, maintained that several large Western medical centres who had received human tissue samples from Russia for the conduct of clinical trials had in fact used the samples to develop 'genetically engineered biological weapons' aimed against the Russian population. According to the report, these weapons were 'capable of rendering Russia's population sterile and even killing it off'. The medical centres allegedly implicated in the plot included the Harvard School of Public Health, the Swedish Karolinska Institute, and the Indian Genome Institute. The report prompted the Russian Federal Customs Service (FTS) to ban export of all human medical biological materials, from hair to blood samples, from Russia (Butrin *et al.* 2007).

The above concerns have been addressed in a number of rules and regulations. Paradoxically, this has resulted in a thicket of diverging and often conflicting regulations which have added layers of complexity. This complexity is further compounded by the number of jurisdictions involved, as evidenced in Chapters 8 (the UK framework), 10 (the Swiss framework), 11 (the French framework), 12 (the Italian framework), and 14 (comparing German and European guidelines) of this volume. Evidently, donor, institutional, and national concerns arise every time tissue is collected and used for research, whether at home or abroad. However, these concerns are amplified when the tissue is shipped across borders, as the stakeholders have no assurance that the country of destination provides 'adequate protection' of their respective interests. This raises the question of how these concerns can be addressed and how adequate safeguards can be given.

13.4 **How to address national concerns**

13.4.1 **No export**

An intuitive and indeed pragmatic solution is to 'just say no' to the export of collected tissue to third countries. When the tissue is kept in the country of origin, any concerns over its abuse overseas will be avoided. In addition to outright export bans, some countries have enacted legislation prohibiting tissue collections from their citizens to be shipped abroad. Article 18 (4) of the Estonian Human Genes Research Act provides that 'all tissue samples (in the GeneBank, JAB) shall be stored in the territory of the Republic of Estonia'. However, the Government of the Republic may, at the request of the chief processor of the collection and if good reasons therefore become evident, grant permission for tissue samples to be stored outside the territory of the republic if it is ensured that the chief processor has effective control over the tissue samples and that the tissue samples cannot be used in a manner prohibited by Estonian law.[3] In Iceland, the transport of human tissue other than for diagnosis or quality control is subject to the approval of the National BioEthics Committee and the Data Protection Authority.[4]

[3] Estonian Human Genes Research Act, passed 13 December 2000, RT I 2000, 104, 685, came into force 8 January 2001.

[4] Icelandic Act on Biobanks, No.110/2000, 12 September 2000.

13.4.2 Export control

With respect to national concerns for bioterrorism, EU Council Regulation No 1334/2000 of 22 June 2000 could apply. This regulation sets up a Community Regime for the control of exports of dual-use items and technology.[5] Dual-use goods and technologies are products and technologies which are normally used for civilian purposes but which may have military applications. Annex IV of the Regulation details the items that are controlled prior to being transferred within the Community. To that end, the Annex implements internationally agreed dual-use controls including the Wassenaar Arrangement, the Missile Technology Control Regime (MTCR), the Nuclear Suppliers' Group (NSG), the Australia Group, and the Chemical Weapons Convention (CWC). As of 28 October 2009, human tissue samples were not listed in the Annex.

13.5 How to address institutional concerns

13.5.1 Control through 'in-house analysis'

Existing biobanks differ in the way that they make their samples accessible. Some, like the UK Biobank, do not release their samples to researchers but instead perform the analyses requested by third-party researchers themselves, unless there is a compelling case for physical release to the researcher. Others encourage third-party requesters, including those from abroad, to use the services of locally designated parties for genotyping rather than genotyping in their own laboratories, for reasons of costs, speed, and creation of added value for the collection.

However, the choice of in-house analysis may be dictated not only by considerations of cost, efficiency, or the prospect of raising extra revenue, but also by considerations of data protection and commitments made to sample donors (e.g. the samples will not leave the country or even the repository). Export bans or in-house analysis policies are likely to alleviate donor concerns. In addition, these concerns may be lessened by informing donors about (possible) international shipment of their samples, the transfer policies to which applications for international shipment will be subject, and the conditions attached to such shipment. Based on this information, donor interest could be further protected by stipulating that donors consent to international transfer of their samples.

13.5.2 Institutional control through access policy

Any institutional challenges could be addressed by establishing a policy for sharing and distributing samples. Such a policy should include a clear and transparent procedure for reviewing requests for international transfer of samples. The policy is to make clear exactly what tissue can be transferred as well as define what tissue will not be transferable. Further safeguards can be built in by establishing a separate transfer committee. The elements to be considered by such a committee when reviewing a request for transfer could include, but are not limited to, the following:

1. the nature of proposal;
2. whether the proposal meets the scientific objectives of the bank;
3. the potential impact of the science;
4. whether the proposal concerns a high- or low-impact disease;
5. whether the requesting party is a for-profit or a not-for-profit organization;

[5] Council Regulation (EC) No 1334/2000 of 22 June 2000 setting up a Community Regime for the control of exports of dual-use items and technology.

6. the credentials and competence of the requesting party and his/her institution;
7. the quality of the proposal;
8. the types of analysis proposed to be performed on the tissue;
9. the amount of tissue requested per analysis;
10. the scope of informed consent;
11. commitments to participants to return results and/or benefits accrued by the proposed research;
12. the willingness of the requesting party to provide reciprocal access on similar terms;
13. the willingness to pay for a return on investment, with the possibility of differentiating fees by type of requester (for-profit or not-for-profit).

13.5.3 Institutional control through transfer agreement

Next to having in place a policy for sharing and distributing samples and a clear and transparent procedure for reviewing requests for transfer of samples, it is considered 'best practice' to only allow the transfer of samples on condition that a material transfer agreement (MTA) has been implemented (this contractual agreement practice is used by French biobanks, for example (see Chapter 11, section 11.2). The implementation of an MTA not only serves to allocate intellectual property rights or to secure proper handling of the material, but also serves as a means by which the tissue bank can meet its obligations towards its participants who donated the material and who may continue to donate samples during the lifespan of the biobank. For example, the MTA could contain a representation on the part of the transferring institution that proper informed donor consent for the transfer has been obtained. The MTA also provides a means of defining and limiting the purposes for which the tissue will be used, for example by allowing or prohibiting genetic characterization of the tissue. A condition which must be met before any tissue can be transferred is that the recipient has the approval of the competent medical ethical review board for the proposed research on the tissue, as set out in the Protocol attached to the agreement. Other conditions could pertain to the provision of evidence by either party that they have obtained all appropriate import and export licenses. Typical undertakings in such MTAs are that the recipient may only use the tissue 'for the common good in scientific research', that the recipient shall not use the tissue for profit or any other commercial gain, and that the recipient may not use the tissue in human subjects. Other undertakings relate to confidentiality (e.g. the recipient shall refrain from tracing or identifying the donor of the tissue), feedback of findings and data (e.g. recipient shall share all data, results, and analyses generated in the course of its processing of the tissue), and ownership and intellectual property (e.g. title to the tissue is and remains in the ownership of the tissue bank).

13.6 How to address donor concerns

13.6.1 Donor control through consent and/or property

The best way for a donor to control international transfer of tissue is via consent (however, see Chapter 2 for the limits of this concept). It seems to follow from the traditional doctrine of informed consent for participation in research that institutions have a legal obligation to inform donors that their tissue might be transferred for research abroad. Consequently, as part of the consent process, donors could withhold their consent to international transfer. Consent for international shipment of samples is routinely asked for in consent documentation for clinical trials.

In tissue research involving volunteers donors are also being informed that their tissue may be shipped across borders and asked for explicit consent to such shipment. In addition to the control afforded by the requirement of informed consent, Dutch donors can exercise control over their tissue, after it has been shipped abroad (with their knowledge and consent), on the basis of their property rights in their tissue, provided that they have not waived such rights. Under the Dutch Act on Conflicts of Law regarding chattels, donor property rights in their tissue will not be affected by the transfer of such property to another state. This could include, for example, a civil law action for 'restitution' of their tissue.

13.6.2 Adequate safeguards

Currently, there is no EU legislation with direct application to the use of human tissue for research. Of course, there is an EU Regulation for the Quality and Safety of Human Tissue for Therapeutic Use in Humans (Transplants), but this Regulation does not apply to the issue at hand; its scope is limited to human tissue that is used as a transplant or implant for therapeutic use in another human. Although European law is silent on the issue, the European Union is bound by the fundamental human rights and freedoms which are laid down in the European Convention for the Protection of Human Rights and Fundamental Freedoms (ECHR) and in the protocols attached to this Convention. Examination of the various protocols of the Council of Europe revealed Recommendation Rec(2006)4 on research on biological materials of human origin. Article 16 of this Recommendation provides that transport of human tissue to third countries is allowed, provided that 'adequate safeguards are in place'.[6] This raises the question of what safeguards can be considered'adequate' and how such safeguards can be put in place. The answer can be found in the Explanatory Notes to the Recommendation, according to which the adequacy of the level of protection afforded by the importing state must be assessed in light of all the circumstances surrounding the transfer. The assessment must consider the nature of the biological materials, the purpose and duration of the proposed research, the state of origin, and the security measures complied with in that state.

13.6.3 Contractual clauses

Notably, the Explanatory Notes to the Recommendation specifically provide that adequate safeguards may result from *contractual clauses*. However, while this Council of Europe Recommendation provides a legal instrument for the legitimate transfer of human tissue from one country to another (i.e. a contract), it stops short of providing both the specifics and the enforcement mechanisms for such a contract. What clauses should such a contract contain in order to satisfy the requirement of 'adequate safeguards'? Who should be the parties to such a contract? And, as contracts only bind those who are privy to them, how could a third party benefit from the safeguards it contains? For the answers to these questions, we could revisit the laws of the European Union, if draw an analogy between the transfer of human tissue and the international transfer of personal data. As we will show, the concept of using contractual clauses to ensure 'adequate safeguards' for cross-border transmission is already deeply rooted in EU data protection law.

6 Article 16 (Transborder Flow) of the Council of Europe *Recommendation Rec(2006)4* (Council of Europe 2006).

13.7 EU law on international transfer of personal data

13.7.1 The principle of adequacy

In the 1990s, the European Union harmonized the various national laws of its member states on the protection of personal data by adopting Directive 95/46/EC (henceforth 'the Directive') on the protection of individuals with regard to the processing of personal data and on the free movement of such data (European Parliament and the Council of Europe 1995). The objective of the Directive is to authorize the free flow of personal data between EU member states by imposing a harmonized set of minimum standards protecting the fundamental right of the citizens of member states to privacy with respect to the processing of personal data (Directive, Article 2, Poullet 2006). In addition to establishing the free flow of personal data within the EU, the Directive also provides a mechanism for the transfer of personal data from an EU member state to a non-EU state. The Directive provides that personal data can only be transferred from an EU member state to a third country if the third country has an adequate level of data protection (Directive, Article 25(1)). The adequacy of protection is to be 'assessed in the light of all the circumstances surrounding a data transfer operation or set of data transfer operations' (Directive, Article 25(2)). The principle of adequacy is not concerned with the general provisions of law in a third country with regard to data protection, but with the actual level of protection which will be afforded in a particular situation (Aldhouse 1999; Kuner 2001). As a result, the determination of an adequate level of protection is based on a case-by-case scenario instead of evaluating the country overall.

13.7.2 Adequacy by contract

Even transfers of personal data to a third country not offering an adequate level of control may still be authorized using various possibilities under the Directive. One such possibility is the condition that the data exporter adduces adequate safeguards (Directive, Articles 26(1) and 26(2)). Specifically, such safeguards may result from appropriate contractual clauses (Directive, Article 26(2)). In other words, under the Directive, transfers of personal data to third countries, which have not yet been determined by the Commission as offering an adequate level of protection, may be authorized through the use of a *contract* to be agreed to by the data exporter and the data importer. To establish what contractual clauses offer an adequate level of protection, the Directive provides that the European Commission decides that certain standard contractual clauses offer sufficient safeguards. The member states must comply with the Commission decision (Directive, Art. 26 (4)).

13.7.3 EU standard contractual clauses

In its published Decision of 27 December 2001,[7] the European Commission adopted a set of standard contractual clauses pursuant to Article 26(2) of the Directive, i.e. clauses that are considered as offering adequate safeguards with respect to the protection of the privacy and the fundamental rights and freedoms of individuals and as regards the exercise of the corresponding rights[8]. The overall idea of using the EU standard contractual clauses is to ensure that data exporters and data importers are assured that the data transfer will have adequate protection (Zinser 2003). In these clauses, the parties agree to a legally enforceable declaration whereby both parties,

[7] Commission Decision 2001/497 of 15 June 2001, on Standard Contractual Clauses for the Transfer of Personal Data to Third Countries under Directive 95/46/EC, 2001 OJ (L 181/19) 1539.

[8] Commission Decision 2001/497, Article1.

the data exporter and the data importer, process the data in compliance with basic data protection rules and principles (Sotto *et al.* 2008).[9]

13.7.4 Content of the EU standard contractual clauses

The EU standard contractual clauses (SCCs) provide an adequate level of protection in third countries by imposing a list of obligations to be satisfied by both the data exporter and the data importer. In general, the data exporter must comply with the laws of the EU and ensure that the data are adequately protected. The data importer, on the other hand, must guarantee that he/she is not aware of any national legislation that prevents him from fulfilling the obligations under the SCCs and that the data are processed appropriately. The obligations include, but are not limited to, the obligation to process the personal data in accordance with the mandatory data protection principles set out in Appendix 2 and to submit its data processing facilities for audit. The principles set out in Appendix 2 include purpose limitations, data quality and security, and the rights of the data subject to access, rectification, deletion, and objection. Although there is no differentiation among the transfer of general personal data and sensitive data under the Directive and the Commission Decisions, the standard contractual clauses require the data exporter to notify the data subject that such sensitive data may be transferred to a country that does not provide an adequate level of protection.[10]

13.7.5 Member state control

While the conclusion of a contract containing the SCCs provides a sound legal basis for the international transfer of the data, it does not, by itself, obviate the need to have a legal basis for making the transfer of data in the first place. This follows from the fact that an international transfer of data is a form of processing of personal data and thus requires a legal basis. Although the SCCs are an instrument to satisfy the requirement of adequacy, the data protection authorities in the member states still retain powers to prohibit or suspend data transfer in exceptional circumstances (ibid.). National data protection authorities may also issue ad hoc contractual clauses to be used in agreements for the transfer of data based on national law (ibid.). However, if the data transfers are made under contracts using the SCCs, member states cannot exercise their powers (ibid.).

13.8 Donor as third-party beneficiary

13.8.1 Third-party beneficiary

While SCCs should go a long way towards alleviating concerns over abuse, as a general matter of law, contracts only bind those who are privy to them. This raises the question of how the protection of individual donors could benefit from contracts between institutions and foreign researchers. The answer is provided by the civil law doctrine of the 'third-party beneficiary'. In fact, this solution has been worked out in the EU regime for SCCs. Analysis of the SCCs reveals another feature which is likely to reinforce the position of the donor. Recital 16 to the Commission Decision[11] provides that the SCCs should be enforceable not only by the organizations which are

[9] In addition to clauses for controller-to-controller transfers, the Commission has also adopted a set of contractual clauses for controller-to-processor transfers of data, Commission Decision 2001/16 of 27 December 2001 under Directive 95/46/EC, 2002 OJ (L 6/52).

[10] Clause 4(b) of Commission Decision 2001/16 of 27 December 2001.

[11] Commission Decision 2001/16 of 27 December 2001.

parties to the contract, but also by the data subjects whose data is being transferred between the two organizations. In particular, this applies where the data subject suffers damages as a consequence of a breach of the contract. The governing law of the contract should be the law of the member state in which the data exporter is established, enabling a third-party beneficiary to enforce a contract on his 'home terms and turf'.

13.8.2 **Enforcement by (associations of) donors**

In Chapter 2, Barilan provides anecdotal evidence to the effect that contracts must be enforceable if they are to avoid violations of human dignity. According to legend, the undertakers who, in consideration of a hefty fee, promised Charles Byrne (the poor Irish Giant) that they would bury him at sea, broke their promise and sold his skeleton to the surgeon John Hunter instead.

Enforcing the terms of the SCCs could indeed be problematic for an individual donor. However, under the EU regime, donors are not left to their own devices. The EU regime provides that data subjects are allowed to be represented by associations or other bodies if they so wish and if authorized by national law. This enables donors who participate in a major biobank to unite and have their representative organization act as a watchdog for the use of their tissue overseas, backed up by the power to initiate legal proceedings on behalf of the donors. Instead of setting up their own organization, donors could opt to have themselves represented by an established patient or consumer advocacy group. To further reduce practical enforcement difficulties, the SCCs stipulate that the data exporter and the data importer should be jointly and severally liable for damages resulting from any violation of those provisions which are covered by the third-party beneficiary clause. Therefore, when data subjects have suffered any damage as a result of the breach of SCCs, by either party, they can take action and receive compensation from the data exporter, the data importer, or both.

Notably, the statutorily imposed mechanism of making data subjects third-party beneficiaries of protective measures, backed up by joint liabilities and accessible remedies, provides a powerful and practical means of helping donors realize the 'key concepts of the ethical debate' discussed in Part I, specifically human dignity, respect for the person, and trust in scientific research.[12] Even more notably, so far the legal–ethical debate of these key concepts has failed to even note the availability of these European safeguards.

13.9 **Combining MTA and EU standard contractual clauses**

The above analysis shows that both institution and donor concerns over cross-border flows of their tissue could be addressed by having an agreement in place: an MTA to address institutional concerns and a contract containing SCCs to address donor concerns. In order to promote efficiency and avoid duplication of bargaining and paperwork, the two contracts should be combined. The combination of MTAs and the SCCs further fosters 'individual as well as collective *empowerment* that would enhance and support public trust and sustainable partnerships in research' (Kanellopoulou in Chapter 5, section 5.6).

Combining MTAs and SCCs also creates the opportunity to make the necessary additions to the SCCs. While the analogy between tissue transfers and transfer of personal data goes a long way, it is not complete. Tissue is different from mere personal data, so when the SCCs are being used as a legal basis for transfer of tissue, appropriate provisions addressing tissue-specific concerns must be added. Such provisions can be found in the standard MTA discussed above and

[12] For an exploration of the notion of trust, see Chapter 5.

might include restrictions on the use of the tissue, the obligation to return unused parts of the tissue, prohibitions of commercial use, etc.

13.10 Conclusion

Concern and uncertainty over the legality and conditions of cross-border tissue flows hampers EU research. It follows from both EU law and the laws of the Council of Europe that cross-border tissue flows are allowed provided that adequate safeguards are in place. The mechanism developed by the EU, i.e. standard contractual clauses, can be used to establish adequate safeguards for the international transmission of personal data. The SCCs further reinforce the position of the donor, as they are enforceable not only by parties to the contract, but also by the data subjects whose data is being transferred between the parties. The clauses also allow the donor to have associations act on his/her behalf and reduce practical enforcement difficulties by holding the data exporter and the data importer jointly and severally liable for damages resulting from any violation of the SCCs. By integrating the SCCs with the terms of the MTA, the rights and obligations of all stakeholders will be established. Last but not least, this solution demonstrates that the alleged lack of European legal certainty and protection can be remedied by making use of the legal mechanisms that the European Union has had in place for over a decade to help its citizens realize key ethical concepts in biomedical research: their human dignity, respect, and trust in research.

References

Aldhouse, F. (1999). The transfer of personal data to third countries under EU Directive 95/46/EC. *International Review of Law, Computers & Technology*, **13**, 75–9.

Butrin, D., Taratuta, Y., and Sborov, A. (2007). *Russia warily eyes human samples.* Kommersant, 30 May. Available at: http://www.kommersant.com/p769777/r_1/Human_medical_biological_materials_export/ (accessed 2 July 2010).

Council of Europe (2006). *Recommendation Rec(2006)4 of the Committee of Ministers to member states on research on biological materials of human origin.* Available at: https://wcd.coe.int/ViewDoc.jsp?id=977859 (accessed 23 December 2009).

European Parliament and the Council of Europe (1995). *Directive 95/46/EC of the European Parliament and of the Council of 24 October 1995 on the protection of individuals with regard to the processing of personal data and on the free movement of such data.* Available at: http://ec.europa.eu/justice_home/fsj/privacy/docs/95-46-ce/dir1995-46_part1_en.pdf (accessed 23 December 2009).

Hwang, W.S. (2005). *Science, special online collection. Controversy—Committee report, response, and background.* Available at: http://www.sciencemag.org/sciext/hwang2005/(accessed 29 October 2009).

Kuner, C. (2001). EU Regulations threaten international data flows, *International Technical Law Review*, **39.**

Poullet, Y. (2006). EU data protection policy. The Directive 95/46/EC: ten years after. *Computer Law & Security Report*, **22**, 206–17.

Sotto, L.J. et al. (2008). European Union Data Protection Directive 95/46/EC – transfer of personal data outside EU standard contractual clauses. In K.P. Cronin and R.N. Weikers, *Data security and privacy law: combating cyberthreats.* Egan, MN: West.

Weisbrot, D. (2004). *Australian Law Reform Commission, essentially yours. Human genetic research databases: issues of privacy and security, overview of comments* Available at: http://www.alrc.gov.au/events/speeches/DW/2004/20040226.pdf (accessed 2 July 2010).

Zinser, A. (2003). The European Commission Decision on standard contractual clauses for the transfer of personal data to third countries: an effective solution? *Chicago–Kent Journal of Intellectual Property*, **3**, 24–120.

Part III

Practices—disciplinary perspectives

Chapter 14

Ethical recommendations for the use of human biological material stored in pathology archives for research purposes

Christoph Brochhausen, Nabila Ahmed, Nicolas Roßricker[1], and C. James Kirkpatrick

14.1 The need for human tissue

Within recent decades, our knowledge of the biological mechanisms in health and disease has undergone tremendous changes. In particular, results from molecular biology and human genetics, as well as new methods, have opened up interesting perspectives for medical science. New findings have led to a better understanding of molecular mechanisms in human development, physiological homeostasis, and pathophysiological processes. As a consequence for medical science, molecular biological knowledge has opened the door for target-oriented therapeutic agents. Three important fields—drug therapy, tissue engineering (see Chapter 15), and regenerative medicine—have been strongly influenced by these developments, which have made a significant input to the optimization of diagnostic and therapeutic strategies.

Regarding modern drug therapy, the improved understanding of molecular mechanisms, especially in carcinogenesis and tumour progression, has become the basis for the development of pharmaceutical agents against specific target molecules which are expressed individually. Examples of such therapeutic targets are epithelial growth factor receptor (EGFR) and vascular endothelial growth factor (VEGF), which are individually expressed in several tumours. These examples illustrate the profound change that has occurred in our conception of drug treatment—away from the treatment of a disease towards a specific therapy of individual conditions, so-called individualized treatment.

Other promising fields in modern medicine are tissue engineering and regenerative medicine, which represent an integrated concept of engineering and life science. The aim of these new disciplines is the replacement of functional tissue to treat major tissue defects due to trauma, cancer, and degenerative or inflammatory processes (Vacanti and Langer 1999). Scaffold-assisted autologous chondrocyte transplantation is a good example of tissue engineering which is already applied in the clinical situation (Brittberg et al. 1994). Furthermore, a combination of biomaterials and cells is used as a pre-colonized vascular graft for implantation in the vascular system (Deutsch et al. 1999). An example of a biomaterial with incorporated growth factor but without pre-seeded cells is found in induced bone regeneration. Biomaterials can be loaded with bone

[1] Parts of this chapter are taken from the MD theses by Nabila Ahmed and Nicolas Roßricker.

morphogenetic protein-2 (BMP-2), an essential growth factor for osteoblast proliferation and differentiation (Kaito et al. 2005).

Finally, the challenge of tissue engineering and regenerative medicine is the promotion of self-repair processes and the replacement of damaged tissue with either the patient's own functional tissue or functionalized autologous cell-scaffold constructs to reduce the amount of foreign material in the body and to regenerate functional organs instead of simply replacing damaged ones by heterologous implants with major limits of function and biocompatibility (Kirkpatrick et al. 2006; Brochhausen et al. 2009).

From the perspective of these challenging fields it is obvious that the traditional concepts of medical treatment are undergoing a fundamental change. Nevertheless, despite intense research efforts, a paradox has become obvious during these developments. Even though, by analysing cell cultures and animal models, basic science aims at opening the door for individualized therapy, tissue engineering, and regenerative medicine, it is increasingly apparent that these findings cannot be completely extrapolated to the pathophysiological situation in humans. In fact, the better we understand pathophysiological mechanisms on the cellular level or in animal models, the more we have to recognize that these models have significant limitations. A good approach to understanding the complex mechanism in situ is to analyse the morphological manifestations within actual tissues. From this point of view, research with human tissue is of pivotal value for gaining a better understanding of the pathophysiological and regenerative processes with respect to possible clinical applications of tissue engineering and regenerative medicine. The potency of tissue research is demonstrated by modern techniques such as genetic fingerprinting. However, these methods may impinge on fundamental personal rights. Thus, even if there is a pressing need for research using human tissue, there are important questions which require an answer.

1. What are the prerequisites for research with human tissue?
2. What are the legal and ethical limitations on research performed with human tissue?
3. What are the infrastructural requirements for the ethical use of human tissue in the clinical setting?

In this chapter we focus on the legal and ethical requirements for the use of human surplus tissue in pathology archives. Practical guidelines for an informed consent procedure are necessary because the scientific use of archived tissue in institutes of pathology takes place in an ill-defined ethical and legal framework (Brochhausen et al. 2005). Pathologists are responsible not only for the selection and availability of tissue samples for a variety of academic purposes such as teaching, quality control, and research, but also for preventing tissue loss and protecting the personal rights of tissue donors.

We will first summarize relevant international and national regulations regarding the scientific use of human tissue. We will then demonstrate the need for informed consent for the scientific use of human tissue specimens and establish requirements for a formalized informed consent process. Finally, we will make some recommendations for the practical implementation of such a consent process in clinical practice (for an exploration on the meaning of informed consent within clinical routines, see Chapter 17).

14.2 **The use of archived tissue**

Since human tissue research is important for transferring potent scientific advances into possible clinical applications for the further optimization of patient treatment, tissue banks are being established not only for human tissue transplants, such as cornea or bone, but also for research purposes (e.g. tumour banks). Another rich source of human tissue for this type of research are the archives in institutes of pathology. The main task of pathology is the pathological–anatomical diagnosis of abnormal features in biopsies, surgical excisions, and amputation specimens as well as in smear specimens and cytological preparations of body fluids. For this purpose, specimens of human tissues or cells which have been removed for diagnostic or treatment purposes are fixed and embedded in paraffin wax before histological analysis. The paraffin block sections are sliced for histomorphological analysis via histochemical and immunohistological staining procedures and, if necessary, via molecular pathological methods and ultrastructural analysis by electron microscopy. Subsequently, these blocks are stored in archives for several years for documentation reasons and for re-evaluation or further analysis. Re-evaluation has become regular practice, especially in tumour patients, with respect to newly established target therapies which offer further therapeutic options, depending on specific parameters which can be detected in the tissue specimens. Thus, modern pathology plays an integrated and active role within the health-care system, in the clinical diagnostic process, the evaluation of prognosis, and the planning of optimal therapy.

However, in addition to being responsible for the care of individual patients, pathology as a scientific discipline has important functions in elucidating the pathogenesis of diseases and the principal morphological reaction patterns of organs and tissues in response to not only to different pathogenetic conditions, but also to therapeutic procedures, therapeutic agents, and interactions with biological or biophysical transplants.

Thus the aim is not only to help individual patients, but also to clarify the principle mechanisms of diseases and tissue reaction.

14.3 **European recommendations**

The European positions for the use of biological samples are based on important legislative documents such as the Convention for the Protection on Human Rights and Fundamental Freedoms (ETS No.5), the Convention on Human Rights and Biomedicine (ETS No. 164), and an additional protocol concerning biomedical research (CETS No. 195). Among the most important principles reflected in these documents are the dignity of humans and respect for the integrity and fundamental freedoms of humans (see also Chapter 3). These fundamental rights also have to be protected with regard to the application of biology and medicine. With respect to the use of biological human material the documents mentioned above give some guidance to the realization of this guarantee in the member states of the European Union. The aim of *Recommendation Rec(2006)4 of the Committee of Ministers to member states on research on biological materials of human beings*, which was adopted in 2006 (Council of Europe 2006), was to achieve greater harmonization in the member states with respect to the maintenance and further realization of human rights and fundamental freedoms relative to the application of biology and medicine. In detail, the recommendation applies to the removal and storage of biological material of human origin in collections and biobanks as well as to its use for research purposes.

In accordance with other legal provisions, the Recommendation clearly states that the protection of the human whose biological materials are removed, stored, or used for research should be of paramount concern. With respect to this important point it is affirmed that particular protection should be given to humans who may be vulnerable in the context of research. Furthermore, the Committee of Ministers is convinced that biomedical research that is contrary to human dignity and human rights should never be carried out. Thus, it is emphasized that the interest and welfare of the human whose biological materials are used in research prevail over the sole interest of society or science. This paragraph clearly demonstrates that personal rights are superior to the interests of society and science, including also the freedom of science. This position is in line with other European guidelines and directives.

In consequence, an important element which should be taken into account is the right of the individual to agree or to refuse to contribute to biomedical research and, further, that no one should be forced to contribute. Since human material should be donated in a spirit of solidarity (for the implications of this concept, see Chapter 6, section 6.3), it is stressed that tissue banks and collections developed on the basis of donations of biological materials should not be monopolized by small groups of researchers. Therefore appropriate transparent governance of biological materials stored for research purposes is particularly important.

The second important element of the Recommendation is a commentary on the importance of medicine and biological science and their contributions to saving life and improving the quality of life. Therefore the need for research with human biological material is delineated.

Furthermore, the Committee of Ministers regards it as a matter of fact that the progress of biomedical science and practice depends on knowledge and discovery which necessitates research on humans and research involving the use of biological materials of human origin. In addition, it is stressed in the Recommendation that such research may often be transdisciplinary and international. For the legitimization of such research the national and international professional standards in the area of biomedical research and the previous deliberations of the Committee of Ministers and the Parliamentary Assembly of the Council of Europe have to be taken into account.

Even if the member states are not legally committed to the Recommendation on research on biological materials of human origin (Council of Europe 2006), it is suggested that their governments adapt their laws and practices to the guidelines of this document and promote the establishment of practice guidelines to ensure compliance with the provisions contained in it. Furthermore, the Secretary General of the Council of Europe is entrusted with the task of transmitting this Recommendation to the governments of states which are not members of the Council of Europe, which have been invited to sign the Convention on Human Rights and Biomedicine, to the European Community, and to international organizations participating in the work of the Council of Europe in the field of bioethics. This important point demonstrates that the Council of Europe recognizes the fact that modern competitive biomedical research is international and interdisciplinary.

Regarding the scientific use of human biological material from pathology archives, it is of special importance that this Recommendation applies not only to research activities in biomedicine involving the removal of human biological material to be stored for research use only, but also to material which has been removed for purposes other than scientific ones. This may also include tissue which has been removed for diagnostic and therapeutic purposes, the remainder of which is archived in pathology institutes. From this point of view the Recommendation offers important insights ito the development of practical guidelines for the scientific use of archived tissue samples. However, the Recommendation does not apply to embryonic and fetal tissue. Thus, if research purposes are being considered, specific regulations are necessary for the use of archived embryonic and fetal tissue which has been removed for anatomical or pathological analysis within the scope of therapeutic and diagnostic applications. The use of this type of material should be clarified, since it offers new perspectives for innovative fields of medicine, especially as regenerative processes recapitulate developmental mechanisms. Furthermore, important insights into angiogenesis, a crucial stage of wound healing and tumour growth, could be obtained by studying angiogenesis in human development. Therefore further understanding of human developmental processes is of special interest for future medical application not only in the field of tissue engineering, but also for tumour therapies.

Since personal rights are recognized as fundamental rights, the identity of human material is an important parameter in the practical regulation of the scientific use of human biological tissues.

In this respect, the Recommendation makes valuable suggestions for the classification of human material: 'identifiable materials are those biological materials which alone or in combination with associated data, allow the identification of the person concerned either directly or through the use of a code'. If the user of the material has access to the code, the material is referred to as 'coded material'. If the user has no access to the code, which is under the control of a third party, the material is referred to as 'linked anonymized material'. Material which, alone or in combination with associated data, does not allow the identification of the person concerned after reasonable effort is referred to as 'non-identifiable biological material'.

In addition to specific suggestions for obtaining and collecting biological materials for research (including biobanks), Article 12 of the Recommendation is also dedicated to residual biological materials. This article clearly defines the principle requirement of appropriate consent to make material which was not originally removed for research purposes available for research activities. It is also stated that, whenever possible, information should be given and consent or authorization requested before the removal of biological materials. If the proposed use of identifiable biological materials in a research project is not within the scope of prior consent given by the person concerned, reasonable efforts should be made to contact that person in order to obtain consent for the proposed use. If contact is not possible with reasonable effort, the biological material should only be used in the research project if an independent evaluation deems that the research addresses an important scientific interest, that the aim of the research could not reasonably be achieved using material for which consent could be obtained, and that there is no evidence that the person concerned had expressly opposed such research use.

The person concerned may freely refuse consent or withdraw consent for identifiable use of biological materials at any time without leading to any form of discrimination against him/her, in particular regarding the right to medical care.

Unlinked anonymous biological material may be used in research provided that such use does not violate any restrictions placed by the person concerned prior to the anonymization of the material. The anonymization should be verified by an appropriate review procedure.

14.4 **German regulations**

German law does not specifically regulate the scientific use of human tissue. Nevertheless, several laws such as the transplantation law, the tissue law, and the law on data safety deal indirectly with relevant legal aspects, especially with respect to the protection of personal rights. However, for a long time it was unclear which type of use was in accordance with these regulations and which represented a violation of personal rights. To harmonize the different legislative rules dealing with this topic, the Bundesärztekammer (Federal Association of Physicians) issued two statements dealing with the scientific use of human tissue: one dedicated to the use of organs and tissues resected for research purposes and another addressing the specific conditions for the use of human tissue obtained from autopsies (Zentrale Ethikkommission bei der Bundesärztekammer 2003a, b).

The principal aim of these two statements was to clarify and provide guidelines for a balance between the protection of personal rights and the constitutional right to freedom of research. In principle, informed consent is required for any scientific use of human tissue. Thus it is recommended that the donor makes a decision about the range of consent. A wide range of options is possible. One possibility is restricted consent justifying only a specific analysis. Alternatively, broad consent can be given for a range of research purposes. However, even in the case of broad consent, future research and methods which were inconceivable at the time that consent was given are not covered (for a qualified vindication of broad consent, see Chapter 2). In cases in which non-anonymized tissue is to be used, a statement that the patient may withdraw consent at any time without any disadvantage regarding further treatment is necessary.

Although informed consent is defined as the principle prerequisite for the scientific use of human tissue, the two recommendations of the Bundesärztekammer identify exceptional conditions under which informed consent is not necessary. Thus, it is stated that, in exceptional cases, after consideration of the scientific aims and the personal rights of the donor, informed consent is not required if the material is no longer necessary for diagnostic procedures and it will not all be used during the research project. However, only anonymized tissue specimens may be used in such cases, which means that only research projects which do not demand personal data from the donors can be performed without informed consent. A further prerequisite for research projects without informed consent is that no individualized genetic analysis will be performed, and that no scientific results which affect the personal interests of the donor or his/her relatives will be generated. Finally, it must be evident that informed consent can only be sought at a disproportionately high cost.

14.5 Suggestions for a formalized informed consent procedure

Based on the national guidelines proposed by the Bundesärztekammer and the European Recommendation on research on biological materials of human origin (Council of Europe 2006), we define the elements necessary for informed consent, particularly with respect to the development of an informed consent form. As a formalized informed consent procedure, we suggest obtaining informed consent for the possible use of tissue samples at the time when the patient is informed about the therapeutic or diagnostic intervention in which tissue samples will be removed for anatomical–pathological analysis. Since the information about the intervention is given a reasonable period in advance of the intervention itself, the patient has time to consider and make a decision about his/her consent for his/her tissue to be used for scientific purposes. Furthermore, information for the intervention should be provided by a physician who is also able to give further information about the circumstances of the tissue sampling, the analysis in the institute of pathology, and the conditions of storage of the material in the archive of the institute of pathology, as well as about the actual scientific investigation itself.

The consent form should consist of two separate parts: the information sheet and the consent form. The information sheet and a copy of the consent form remain with the person concerned. Thus the person possesses a document about the consent procedure which provides the opportunity for reflection about their decision and offers the possibility of withdrawing their the consent. It is important to note that the information and consent form for scientific use are separate from the consent form for the medical intervention itself. Thus, the consent for scientific use is emphasized, and it is clear that the intervention and the scientific use of the leftover tissue are for entirely different purposes.

In the information sheet it should be explained to the person concerned that the removed tissue will be brought to the institute of pathology for the anatomical–pathological diagnosis, which will then guide further procedures and/or possible diagnostic alternatives. It should also be explained that the remaining material will be stored in the archive of the institute of pathology for documentation purposes as well as to leave open the possibility of re-evaluation and further analysis in future, if necessary, but that it will not generally all be required for these diagnostic procedures. In this context, it should also be emphasized that, after a certain period of time has elapsed, any remaining material will be destroyed.

The information sheet should point out that this material could be of enormous value not only for education and quality management purposes, but also for research in order to gain better understanding of disease mechanisms and to improveunderstanding of possible therapeutic targets. Therefore permission to include parts of the removed tissue in research projects is sought. To give a better impression of the circumstances and extent of the scientific use, it is explained that structural analyses and immunohistological techniques to detect proteins as well as molecular–pathological analyses to detect specific genes or gene products will be performed. Moreover, it is stated that no individualized genetic analysis will be performed. In addition, it is stressed that consent would only be valid for such methods and research projects that are planned at the time when consent is given. New informed consent has to be obtained for future projects using analyses or processing methods, especially those which influence individuality, personal rights, and fundamental freedoms, which had not been developed at the time when consent was originally given. This limitation of consent should be explained in the information sheet. Furthermore, an assurance is given that the material would only be used for projects which had obtained prior approval from the local ethical review board.

It should also be made clear in the information sheet that the material will only be used after the diagnostic process is completed, and that not all the archived tissue will be used for research, so that future diagnostic and therapeutic examination will still be possible if necessary. A further important point is explained that no more tissue than is necessary for the diagnostic or therapeutic procedure will be obtained during the intervention. It is also stressed that the material will not be used for research projects in which any third party has a financial interest.

It should also be explained that the persons concerned can limit the range of consent according to either pre-formulated choices given in the consent form or individual choices. Finally, persons will be informed about their free choice to give consent or to refuse without any discrimination, particularly concerning their medical treatment, and their right to withdraw informed consent at any time without giving any reasons.

The possibility of defining the range of consent is given in the consent form. This means that the patient may exclude specific methods (e.g. molecular biological analyses) or a specific type of anonymization. Since the material will be identifiable, patients may choose to have their samples stored as 'coded material', 'linked anonymized material', or both. In addition, it will be possible to draw up individual agreements concerning the range of consent. Furthermore, there is space on the consent form to document specific issues discussed during the consent procedure. Finally, the patient confirms that he/she has fully understood the circumstances and the scope of the scientific use of the material. The consent form is then signed by the patient and the physician who provided the information. One copy of the consent form is attached to the patient's record; another copy and the information sheet are kept by the patient. A copy of the consent form should also be attached to the record system of the institute of pathology to permit adherence to the agreed range of options. Furthermore, the paraffin blocks designated for scientific purposes should be clearly marked. However, every use of the blocks should be documented in a databank to ensure that no material is lost.

14.6 **Conclusion**

In this chapterer, we have demonstrated the need for tissue research. Furthermore, we have argued that not only biobanks but also pathology archives are of value for tissue-related research. However, practical guidelines for the use of this tissue are necessary, and relevant aspects of these guidelines can be found in national and international judicial documents. An ethical framework delineating the extent and prerequisites of the use of human tissue for research is available in European and German ethical regulations. As a fundamental principle, German and European regulations state that, whenever possible, information should be given and consent or authorization should be requested before human biological material is removed. Based on these guidelines, we propose elements for a formalized informed consent process for the scientific use of human biological material from pathology archives. This includes information about the scientific use of the material. It should also allow the patient to limit the extent of use for research purposes. Furthermore, we suggest an infrastructure which guarantees the proper use of tissue for which informed consent was given, as well as the exclusion of tissue from research for which consent was not given or for which consent has been withdrawn. Finally, we suggest a data management process to ensure that adequate material is retained for re-evaluation for diagnostic or therapeutic strategies.

In conclusion, we hope that our suggestions offer an unambiguous, transparent, and practical approach to the proper use of human tissue samples from pathology archives in research projects. The proposed informed consent process is based on a balancing of individual rights against society's need for tissue research and freedom of research (for further suggestions on how to balance interests in the field of human tissue research, see Chapter 5). Thus our recommendations will help to clarify the ethical use of human tissue in research projects. Nevertheless, more deliberation is required on the use of material that is already in storage. A further important cionsideration is how society can be adequately informed about the storage of human material and the enormous treasure which this material represents for the further optimization of diagnostic and therapeutic strategies. In this discussion, pathologists should actively engage not only in explaining possible options, but also in reducing fears concerning the protection of personal rights in human tissue research.

References

Brittberg, M., Lindahl, A., Nilsson, A., Ohlsson, C., Isaksson, O., and Peterson, L. (1994). Treatment of deep cartilage defects in the knee with autologous chondrocyte transplantation. *New England Journal of Medicine*, **331**, 889–95.

Brochhausen, C., Rossricker, N., Wolfslast, G., Weinrich, C., and Kirkpatrick, C.J. (2005). The use of archived tissue—comparison of German and British regulations. *Pathology, Research & Practice*, **201**, 200.

Brochhausen, C., Lehmann, M., Halstenberg, S., Meurer, A., Klaus, G., and Kirkpatrick, C.J. (2009). Growth plate cartilage as an innovative developmental model to target signalling molecules and growth factors for the tissue engineering of cartilage. *Journal of Tissue Engineering and Regenerative Medicine*, **3**, 416–29.

Council of Europe (2006). *Recommendation Rec(2006)4 of the Committee of Ministers to member states on research on biological materials of human origin.* Available at: https://wcd.coe.int/ViewDoc.jsp?id=977859&Site=COE&BackColorInternet=DBDCF2&BackColorIntranet=FDC864&BackColorLogged=FDC864 (accessed 05 January 2010).

Deutsch, M., Meinhart, J., Fischlein, T., Preiss, P., and Zilla, P. (1999). Clinical autologous in vitro endothelialization of infrainguinal ePTFE grafts in 100 patients: a 9-year experience. *Surgery*, **126**, 847–55.

Kaito, T., Myoui, A., Takaoka, K. et al. (2005). Potentiation of the activity of bone morphogenetic protein-2 in bone regeneration by a PLA-PEG/hydroxyapatite composite. *Biomaterials*, **26**, 73–9.

Kirkpatrick, C.J., Fuchs, S., Peters, K., Brochhausen, C., Hermanns, M.I., and Unger, R.E. (2006). Visions for regenerative medicine: interface between scientific fact and science fiction. *International Journal of Artificial Organs*, **30**, 822–7.

Vacanti, J.P. and Langer, R. (1999). Tissue engineering: the design and fabrication of living replacement devices for surgical reconstruction and transplantation. *Lancet*, **354**, 132–4.

Zentrale Ethikkommission bei der Bundesärztekammer (2003a). Die (Weiter-) Verwendung von menschlichen Körpermaterialien für Zwecke medizinischer Forschung (The use of human biological material for medical research). *Deutsches Ärzteblatt*, **100**, A1632.

Zentrale Ethikkommission bei der Bundesärztekammer (2003b). Die (Weiter-) Verwendung von menschlichen Körpermaterialien von Verstorbenen für Zwecke medizinischer Forschung (The use of human biological material from autopsies for medical research). *Deutsches Ärzteblatt*, **100**, A2251.

Chapter 15

Informed consent when donating cells for the production of human tissue engineered products

Leen Trommelmans, Joseph Selling, and Kris Dierickx

15.1 Introduction

Tissue engineering (TE) can be defined as 'the creation of new tissue for the therapeutic reconstruction of the human body, by the deliberate and controlled stimulation of selected target cells, through a systematic combination of molecular and mechanical signals' (Williams 2006). The in vitro combination of donated, cultured, and substantially altered human cells with supporting structures (scaffolds) and biomolecules such as growth factors yields metabolically active constructs, known as human tissue engineered products (HTEPs) (Trommelmans et al. 2007).

Donated cells are the essential constituents of HTEPs and are derived from various sources, including human embryonic stem cells, adult stem cells, and differentiated cells (Guillot et al. 2007). Donation requires the informed consent of prospective donors (Council of Europe 2002). They need to know how donation is going to affect them, its advantages, and its risks. These risks are not limited to the potential physical harm to donors, but also include non-physical harm (for an assessment of the risks involved, see Chapter 4, section 4.2). One may harm donors non-physically by interfering in their autonomy, privacy, or personal integrity (e.g. by using the donation for research purposes to which the donor strongly objects) (Eriksson and Helgesson 2005), or by exploiting cells in a commercial environment without the donor's knowledge (Andrews and Nelkin 2001). As cell donation should not be presumed or obliged, potential donors have to be persuaded to donate (different approaches for encouraging donor participation are outlined in Chapter 6, section 6.2). Donation will be more likely if one can demonstrate that it is in the persons' best interest and/or if donation is consistent with their world view (Wendler and Grady 2008). Therefore donors have to receive adequate information on the whys and hows, not only of the donation itself, but also, as we will explore, the further use of their cells.

Central to the discussion of what constitutes adequate information in a TE context should be the consciousness that cells are not mere biological entities but are laden with various concepts of value. Stakeholders may use the term 'value' in diverse, sometimes conflicting, senses and derive different ethical precepts from them. Therefore awareness of these meanings should precede the question of what should be addressed in the informed consent procedure.

We restrict ourselves to cell donation by living competent adult persons for the purpose of creating HTEPs in a European context. Other ways of procuring cells are beyond the scope of this chapter as they may raise additional problems.

First, we situate cell donation within the framework of TE and its regulation. Secondly, we focus on some of the features of cell donation for TE. Finally, we examine the different notions of 'value' that can be attached to cells—biological, relational, financial, and informational—and

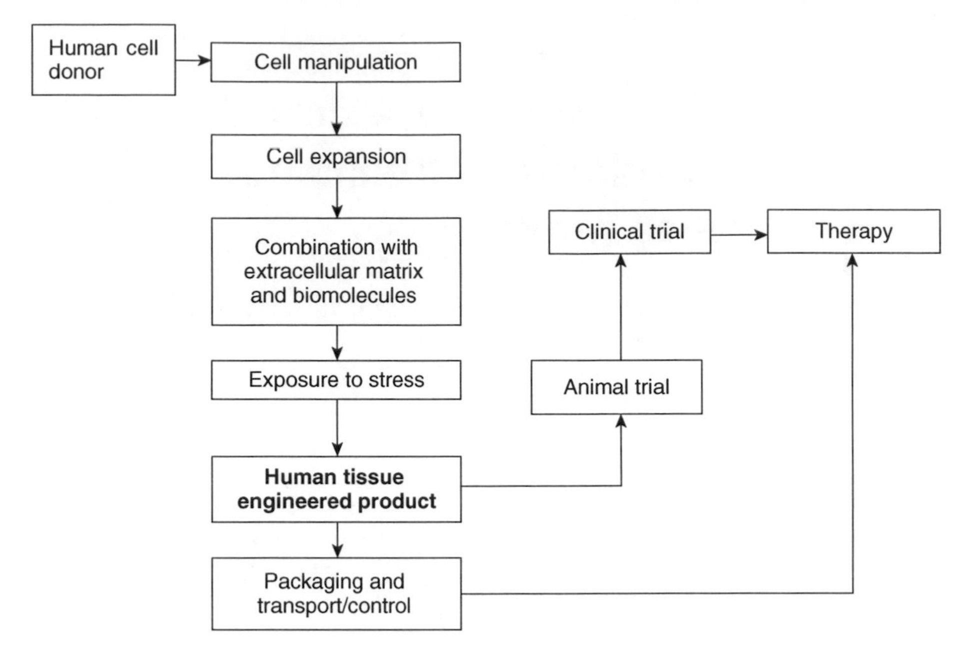

Fig. 15.1 The tissue engineering process.

how these notions may influence the organization of cell transfer, the informed consent procedure, and the contents of the informed consent form.

15.2 **Scope of tissue engineering**

Definitions of TE (Williams 2006; European Parliament and Council of Europe 2007) suggest that HTEPs will only be applied for therapeutic purposes. However, cell donation is also essential for fundamental TE research, for developing human tissue and disease models (Griffith and Swartz 2006; Kirkpatrick et al. 2007), and for applications without any therapeutic benefit including alternatives to animal testing (Poumay and Coquette 2007; Balharry et al. 2008), development of sensors (MATES 2007), and even the creation of jewellery (Anon. 2006). From the donors' perspective, the donation is the same, regardless of the later use of the cells, yet this actual use may influence their decision (not) to donate as some TE applications may touch on deep-seated values or fears (see Chapter 13, section 13.3). Animal rights supporters might be willing to donate for the development of tissue models replacing animal studies, while others might want to donate for therapeutic purposes only, or may want to receive a financial reward for their donation for a non-therapeutic TE application. Therefore presenting TE as restricted to therapeutic applications is misleading. Consequently, a core issue of the informed consent procedure would be provision of information to prospective donors on the foreseeable use of the HTEP, allowing them to decide to donate for a specific goal (therapeutic/non-therapeutic type of HTEP). If this information cannot be provided, the various possibilities (as far as known) should be pointed out (Schulz-Baldes et al. 2007).

15.3 **Regulation of tissue engineering**

For therapeutic goals, obtaining informed consent is mandatory. EU Directive 2004/23/EC (henceforth 'the Directive') stipulates the modalities for tissue and cell transfer (European Parliament and Council of Europe 2004). However, regulations must be critically re-examined whenever new technologies emerge to check whether the fundamental values that have inspired these regulations are still served in the best possible way.

Informed consent aims to protect the autonomy and dignity of donors. The annex to the Directive stipulates that they must be informed:

> about the purpose and nature of the procurement, its consequences and risks, analytical tests if they are performed; recording and protection of donor data, medical confidentiality; therapeutic purpose and potential benefits and information on the applicable safeguards intended to protect the donor. The donor also has the right to receive the confirmed results of the analytical tests, clearly explained.

Here only the relationship between donor and immediate procurer is addressed; whatever happens afterwards lies beyond the Directive's governance, even though this may be important for the donor. Also, the Directive is limited to human applications and manufactured products derived from human tissues and cells *and* intended for human applications (Article 2, section 2). Cell donation for other purposes is not covered.

Cell traceability from donor to recipient and vice versa is enforced. The regulation on HTEPs equally requires complete traceability of the HTEP, its components, and sources, and of the patient.

The production of HTEPs as advanced therapy medicinal products is regulated by EU regulation (EC) No. 1394/2007 (European Parliament and Council of Europe 2007). These products are also defined as 'for human use'. However, some TE applications clearly do not seek any therapeutic benefit, not even broadly speaking, as we argued earlier. These HTEPs may not be covered by either regulation, but they still require donated cells. This creates two donor groups: those who are protected by the Directive and those who are not. If the Directive covered all cell donations, all donors would be treated equally.

15.4 **Features of tissue engineering**

Cells have been donated, stored, altered, and transplanted for many purposes (Waldby and Mitchell 2006; Landecker 2007). Numerous publications have been dedicated to the ethical issues of cell donation. How, then, can TE influence these issues? We discern three elements.

1. Although implanting a HTEP bears some resemblance to 'normal' tissue or organ implantation, one characteristic of TE is fundamentally different: donated cells are not transferred directly from donor to recipient, but are considerably altered, and the in vitro construction of the HTEP creates a markedly more complex product (Trommelmans 2007).
2. TE requires intermediaries involved in the production of the HTEP: cell-banking institutions, TE facilities, or others. The HTEP may be implanted in a person other than the donor. Thus, providing cells for therapeutic allogeneic HTEPs generates a network of relationships where at least three parties are 'entangled': the donor, the intermediary/ies, and the recipient, each with specific interests. The connecting factor is the donated cell (Waldby and Mitchell 2006). Even when cells are not used for therapy, various stakeholders are involved in the production (Trommelmans 2007).

3. Cell traceability (European Parliament and Council of Europe 2007) ensures that the link between donor and donated cells is perpetuated throughout the TE process. Therefore if donors want to protect their interests in the donated cells, traceability makes it possible.

Having active cells derived from one donor and functioning in three environments, i.e. the donor, the HTEP under construction, and the recipient, creates a new reality. Cells in TE are not just biological matter—they establish relationships. Secondly, cells can be donated or sold, placing them firmly in tissue economies (Waldby and Mitchell 2006). Finally, cells are repositories of (genetic) information which connects them with the person from whom they have been derived.

15.5 **Value of cells**

TE forms the backdrop of our understanding of the various values associated with cells. The *Oxford English Dictionary* (*OED*) gives various definitions of 'value' which are, albeit often implicitly, applied to cells. However, some of these definitions entail mutually exclusive consequences.

1. For the tissue engineer, the biological value of cells lies in 'their power and validity for a specific purpose or effect' (ibid.), i.e. TE.
2. However, the value of bodily material can also refer to 'the relative status of a thing, or the estimate in which it is held. In philosophy and social sciences regarded esp. in relation to an individual or group' (ibid.). Here, cells are valued for what they represent: the human being and the relationships they establish. Touching on cells implies touching on the human being, and the respect we owe it.
3. A third definition refers to 'the material or monetary worth of a thing' (ibid.). Therefore cells are also entities that are exchangeable for money.
4. We would confer here a fourth type of value to human cells: their value as a result of the (genetic) information they contain.[1]

Stakeholders in TE may have widely differing, sometimes conflicting, views and interests derived from these values, based on strong ethical convictions and/or economic interests (potential ways of balancing the interests involved in human tissue research are addressed in Chapter). The donor must be able to discern these values and how they influence the decision to donate and the contents of the informed consent.

15.5.1 **The biological value of cells in TE**

Cells are indispensable for TE. Deriving and using them efficiently and safely are prime concerns (Gardner 2007). Safety concerns will guide cell and donor selection (e.g. based on their HLA type). Cells with a specific HLA type may be more coveted than others, given the need to match the cell type of the HTEP with that of the recipient to avoid detrimental immune responses. Consequently, all donors are valuable, but some may be more valuable than others (Bok et al. 2004). This in turn may influence how potential donors value their cells. For them, not the biological value in itself matters, but how it can generate other types of value (see Lenk in this volume).

[1] This fact might provide a justification for giving feedback on health-relevant findings to donors (see Chapter 6).

15.5.2 The relational value of cells

People may find it in their best interests to convert their biological riches into financial value, but whether we allow individuals to exploit their inherited biological resources depends on how we appreciate the relational and material value of their cells (Björkman 2006; Schulz-Baldes et al. 2007).

A relational value has always been assigned to bodily material. Recital 18 of the Directive is exemplary: cell donation should be regarded as an act of altruism, promoting social coherence, solidarity between donor and recipient, and contributing to the general welfare: citizens are encouraged to donate cells under the theme 'we are all potential donors'. It establishes voluntary and unpaid donation of tissues and cells and advocates the procurement of the tissues and cells *as such* on a non-profit basis. Bodily material is the means through which a social contract is established, linking citizens through a 'gift relationship' (Titmuss 1997). Donation schemes create a (virtual) 'pool' of cells or organs that may one day benefit donors. Therefore it is in their own interest to contribute to the pool, thus increasing the chances of having access to suitable cells in case of need. Such a system may endeavour to include as many cell types as possible, to help as many recipients as possible. However, so far no system can ensure that all persons will find suitable cells (Faden et al. 2003). These public banks figure as intermediaries between donor and TE facility. This may have repercussions for the informed consent procedure.

Alternatively, TE facilities could select donors, creating a restricted pool of donors whose cells comply with the requirements of the TE process. This establishes a close relationship between donors and TE facility, in which the donation's societal dimension recedes to the background. Partners and values attributed to cells have changed: donation is no longer a gift to an institution representing all potential beneficiaries; but a transfer of material to a specific stakeholder whose primary aim is to produce an economically viable product (Cosh et al. 2007). The HTEP may then not necessarily be available to everyone or address the most urgent needs. It also means that donors may look at their cells as a renewable source that has value to a particular recipient, and not as a pointer to his/her body or person. Donor and recipient enter into a relationship governed by supply and demand, where the donor may want to obtain some form of 'return on investment'. In this construction one may use 'cell transfer' instead of gift, given the connotations of altruism attached to 'gift' as it is used in European documents. This option is grounded in the idea that cells are a valuable raw material at the disposition of the individual that can be transacted (Hayden 2007). This is contrary to current prohibitions on commercializing (part of) the body, but in tune with current developments in cell-based technologies (Waldby and Mitchell 2006).

The preferred setting for cell donation will be influenced by the adopted connotations of value and will in its turn influence the preferred relationships: either to strenghten the societal framework, or to establish a privatized link between cell provider and HTEP producer.[2] The relational value engendered by the two routes of cell procurement can be in line with or contrary to the views of prospective donors on medical research and health care. Therefore the aims and scope of the facility should be clearly explained: whether it is a for-profit or a not-for-profit establishment, and whether it processes the donated cells itself or acts as an intermediary. In the latter case donors may want to be informed to whom cells are transferred, and for which aim—commercial or not, therapeutic or other (Wilkinson 2005). If donors cannot be informed about further use, a steward monitoring the use of cells on behalf of donors and/or society as a whole may have to be installed (Schulz-Baldes 2007).

---

[2] For an alignment of individual and societal framing, see the Swedish and Norwegian regulatory approaches for human tissue research, as discussed in Chapter 7.

15.5.3 The (potential) financial value of donated cells

The primary meaning of the informed consent process and the consent form is to respect the autonomy of the donor (for a critical appraisal, see Chapter 2). We argue that informed consent in cell donation takes on a second, implicit, role: as it is the only written document mandatorily accompanying cell transfer, in a certain sense it also serves as a contract, 'a writing in which the terms of the bargain are included' (*OED*). Current European regulations are straightforward about the exact terms of the 'contract': any form of payment is prohibited (Council of Europe 2002; European Parliament and Council of Europe 2004). Countries with a common law tradition and with a civil law tradition prohibit the commercialization of bodily material. In civil law tradition, the human body and its parts are the locus of absolute value and dignity. Commercializing (parts of) the body is assigning a market value to it, and consequently destroying its dignity. Common law countries argue that one has no property rights in and cannot exert ownership over one's body or its parts (Dickenson 2007). However, HTEPs will be produced in an environment where products and/or processes are subjected to intellectual property rights and where the production of a specific HTEP will be governed not only by therapeutic necessities, but also by commercial interests and strategies (Cosh et al. 2007; Wadman 2008). The advance of medicine has de facto made the body an 'open source of free biological material for commercial use' (Dickenson 2007), putting the issue of ownership in a new perspective (see Chapter 9). Currently, the leap from donated material without property status to commercializable material with property status is made through an appeal to a Lockean interpretation of property: HTEPs obtain property status through the admixture of labour with the donated cells (Dickenson 2007). Thus the (potential) financial value of the cells changes radically from one moment to another—from non-existent at donation to enormous once they are cultured (Waldby and Mitchell 2006).

The European prohibition on commercializing donations of bodily material leaves donors with empty hands because their donation is a cession rather than a gift: the donors' autonomy is the autonomy to forgo all further claims to financial return.

There is growing unease concerning the discrepancy between the prohibition to sell one's cells, based on one definition of the value of cells and the resulting high morality of altruism and citizenship, and the incorporation of these same cells in a system that is focused on the commercialization of highly innovative products such as HTEPs, which may not even serve a medical purpose. This discrepancy becomes acute when the potential financial value of the cells increases, they are isolated from a physiologically renewable source, without harm to the donor, and the resulting HTEPs do not serve a medical purpose. If the body is turned into an inexhaustable mine for raw materials, should not the mine-owner, or the steward of the mine, have some part of its proceeds (Nelkin and Andrews 1998; Wilkinson 2005)? Donating renewable cells for a skin-mimicking construct for cosmetic tests hardly compares with a kidney donation for someone with a life-threatening condition. In the first example, the donors' dignity may be harmed more by not receiving a reward for their donation than by establishing a well-regulated regime of payment or compensation. This donation is clearly not an altruistic act towards someone in need. In the second example, asking someone to part from a vital organ in return for even a large amount of money exposes that person to exploitation and may harm his/her health and diminish his/her dignity. TE confronts us with the limits of the rules governing the exchange of bodily material. By putting all cell, tissue, and organ donations in the same basket, some of the quintessential differences between the various types of donations, and part of the context of donations, are obscured, although they are relevant to the fundamental issues at stake—the dignity and autonomy of the donor.

Discussing property and property rights in TE may be a complex issue for other reasons, as shown in the following two examples.

1. Who exerts property rights over an autologous HTEP created from the patient's own cells? The patient or the TE facility that has invested labour in the production of the HTEP and thus has given it its property status? If the TE facility is the owner, can it sell the HTEP to a third party, for example because that party is willing to pay more? Or is creating an autologous HTEP a service and an HTEP not a product?
2. If an HTEP obtains property status following the Lockean interpretation, the recipient 'buys' a product that is the TE facility's property. However, as time progresses, the HTEP should integrate in the body through the labour exerted by the body on the HTEP. Can we then say that, because the recipient's body has added labour to the product, it becomes the owner of the HTEP? Of course, this goes against a basic assumption for prohibiting commodification and commercialization—that one cannot own one's body. Or is the physiological labour of the body no labour (Dickenson 2007)?

If we want to guide cell-sourcing activities in the context of TE and preserve the dignity and interests of donors, concepts such as donation, gift, and property have to be refined.

15.5.4 The informational value of cells

The informed consent form can bar cell donors from having any say in what happens to their cells and from any financial interest. However, donors maintain an interest in the information that is contained in the donated cells that have become part of the HTEP. What is learned from cell behaviour during the production of the HTEP or after its implantation may be valuable for the donor's future health (see Chapter 6).

Because European regulation enforces cell traceability, the link between donor and donated cells is perpetuated (European Parliament and the Council of Europe 2007). Cartilage cells are already assessed for their suitability to be re-implanted by their response to specific markers (De Bari et al. 2008). A similar procedure may be used at a later stage to assess the quality of cells before they are incorporated in allogeneic HTEPs. This information is not only valuable for tissue engineers and for recipients who want to receive well-functioning HTEPs, but is also important for donors; signs of malfunction under laboratory conditions may be indicative of the cells' behaviour in the donor's body. Some 'malfunctions' may be due to the artificial circumstances of the in vitro culturing; but one should not rule out the possibility that what happens in vitro or in the recipient may also occur in the donor.

Donors should be made aware that relevant information may come to light during the TE process and implanting, and should be given the possibility, respecting the recipient's anonymity, to obtain validated information that is relevant for them. As very few donors may be involved in one specific application of TE (Parenteau 1999), this may be practicable. Therefore the consent form should provide information on how donors can be recontacted, if they so wish. Although retracing and informing donors of relevant findings may be a burden for the tissue engineers, this possibility should not be rejected. Showing respect for persons who are providing you with vital cells by giving them crucial information that may lead to their well-being is a commendable mode of action, if not a moral duty. Returning information could also be regarded as a form of benefit-sharing (Schroeder 2007). It would not be very consistent to aspire to the healing of one person through TE while withholding vital information from another person who has significantly contributed to your research and who is traceable.

15.6 Conclusion

TE is a complex process, of which cell donation is a vital part. In discussing the scope and content of informed consent by cell donors, the various connotations of 'value' of cells and the relationships that are established through cell donation must be considered. Therefore provision of information should not be restricted to the donation process or the biological quality of the donated cells. In cell donation, the informed consent form serves a double function—providing adequate information and defining the exact conditions for cell transfer, and thereby taking on some characteristics of a contract. Donors are informed about their rights, or the absence thereof, regarding their cells and the products derived from them. If donors have no say in the specific use of their cells and are barred from financial gains, donation becomes a cession.

Because of the symbolic and relational value that they attach to cells, donors may have valid interests in what happens to their cells after donation and therefore may want to be informed about this. Donated cells may generate a substantial financial value when they are incorporated in HTEPs, and consequently donors may also attach financial value to their cells. Which rights we grant donors depends partly on which type of cell value society wants to promote. Under current European regulation donors are barred from any financial benefit; only the relational and symbolic value of human cells is considered.

Cell donation establishes a network of relationships. The two conflicting types of value also influence the relationships that are privileged. If one emphasizes the cells' relational and symbolical value, a system of public cell banks is the obvious setting, with these banks acting as the cell stewards. Thus the altruistic character of free cell donation is continued in the stewardship of the receiving organization that has a firm societal basis and can act for the common good. However, if a more private relationship between donor and recipient is established, the potential financial value of the cells may become more important. Tensions will probably arise because commercialized HTEPs definitely have financial potential, yet donors are not allowed to have any part in this value. Further research needs to be done in order to clarify the meanings and repercussions of the concept of value and of property of human somatic cells, and the associated rights of donors.

Finally, one can exclude donors from any further say in the use of their cells and from any financial return. However, the information that donors and their donated cells share ensures that the relationship between all involved in the TE process is maintained. The 'disentanglement' of donors, their cells, and the other stakeholders in the TE process is never fully possible, neither is it desirable. We advise that the informed consent procedure should take this into account and allow donors to be recontacted if they so wish.

Further research should be performed to develop a suitable informed consent procedure to protect donors' fundamental rights and to safeguard the ethical principles that govern current regulation, i.e. dignity and autonomy.

Acknowledgements

The authors acknowledge the financial support of the European Commission under the DISC REGENERATION project (NMP3-LA-2008-213904).

References

Andrews, L. and Nelkin, D. (2001). *Body bazaar. The market for human tissue in the biotechnology age.* New York: Crown.

Anon. (2006). *Biojewellery: promoting tissue engineering to the public.* Available at: http://www.biojewellery.com/project.html (accessed 7 September 2009).

Balharry, D., Sexton, K., and BéruBé, K.A. (2008). An in vitro approach to assess the toxicity of inhaled tobacco smoke components: nicotine, cadmium, formaldehyde and urethane. *Toxicology*, **244**, 66–76.

Björkman, B. (2006). Why we are not allowed to sell that which we are encouraged to donate. *Cambridge Quarterly of Healthcare Ethics*, **15**, 60–70.

Bok, H., Schill, K.E., and Faden, R.R. (2004). Justice, ethnicity, and stem-cell banks. *Lancet*, **364**, 118–21.

Cosh, E. Girling, A., Lilford, L., et al. (2007). Investing in new medical technologies: a decision framework. *Journal of Commercial Biotechnology*, **13**, 263–71.

Council of Europe (2002). *Additional Protocol to the Convention on Human Rights and Biomedicine, on Transplantation of Organs and Tissues of Human Origin*. Strasbourg, Report ETS No. 186. Available at: http://conventions.coe.int/Treaty/EN/Treaties/Html/186.htm (accessed 7 September 2009).

De Bari, C., Dell'Accio, F., Karystinou, A., et al. (2008). A biomarker-based mathematical model to predict bone-forming potency of human synovial and periosteal mesenchymal stem cells. *Arthritis and Rheumatism*, **58**, 240–50.

Dickenson, D. (2007). *Property in the body: feminist perspectives*. Cambridge: Cambridge University Press.

Eriksson, S. and Helgesson, G. (2005). Potential harms, anonymization, and the right to withdraw consent to biobank research. *European Journal of Human Genetics*, **13**, 1071–6.

European Parliament and Council of Europe (2004). *Directive 2004/23/EC of the European Parliament and of the Council of 31 March 2004 on setting standards of quality and safety for the donation, procurement, testing, processing, preservation, storage and distribution of human tissues and cells, Official Journal L 102, 07/04/2004 P. 0048 - 0058*. Available at: http://eur-lex.europa.eu/LexUriServ/LexUriServ.do?uri=CELEX:32004L0023:EN:HTML (accessed 7 September 2009).

European Parliament and Council of Europe (2007). *Regulation (EC) No 1394/2007 of the European Parliament and of the Council of 13 November 2007 on advanced therapy medicinal products and amending Directive 2001/83/EC and Regulation (EC) No 726/2004 (Text with EEA relevance), (EC) No 1394/2007*. Available at: http://ec.europa.eu/enterprise/pharmaceuticals/eudralex/vol-1/reg_2007_1394/reg_2007_1394_en.pdf (accessed 7 September 2009).

Faden, R.R., Dawson, L., Bateman-House, A.S., et al. (2003). Public stem cell banks: considerations of justice in stem cell research and therapy. *Hastings Center Report*, **33**, 13–27.

Gardner, R.L. (2007). Stem cells and regenerative medicine: principles, prospects and problems. *Comptes Rendus Biologies*, **330**, 465–73.

Griffith, L.G. and Swartz, M.A. (2006). Capturing complex 3D tissue physiology in vitro. *Nature Reviews: Molecular Cell Biology*, **7**, 211–24.

Guillot, P.V., Cui,W., Fisk, N.M., and Polak, D.J. (2007). Stem cell differentiation and expansion for clinical applications of tissue engineering. *Journal of Cellular and Molecular Medicine*, **11**, 935–44.

Hayden, C. (2007). Taking as giving: bioscience, exchange, and the politics of benefit-sharing. *Social Studies of Science*, **37**, 729–58.

Kirkpatrick, J.C., Fuchs, S., Hermanns, I.M., Peters, K., and Unger, R.E. (2007). Cell culture models of higher complexity in tissue engineering and regenerative medicine. *Biomaterials*, **28**, 5193–8.

Landecker, H. (2007). *Culturing life. How cells became technologies*. Cambridge, MA: Harvard University Press.

MATES (Multi-Agency Tissue Engineering Science) Interagency Working Group (IWG) (2007). *Advancing tissue science and engineering. A foundation for the future. A multi-agency strategic plan*. Available at: http://www.ostp.gov/pdf/advancing_tissue_science-engineering.pdf (accessed 7 September 2009).

Nelkin, D and Andrews, L. (1998). Homo economicus: commercialization of body tissue in the age of biotechnology. *Hastings Center Reports*, **28**, 30–9.

Parenteau, N. (1999). Skin: the first tissue-engineered products. *Scientific American*, **280**, 83–4.

Poumay, Y. and Coquette, A. (2007). Modelling the human epidermis in vitro: tools for basic and applied research. *Archives of Dermatological Research*, **298**, 361–9.

Schroeder, D. (2007). Benefit-sharing: it's time for a definition. *Journal of Medical Ethics*, **33**, 205–9.

Schulz-Baldes, A., Biller-Andorno, N., and Andorno, A.M. (2007). International perspectives on the ethics and regulation of human cell and tissue transplantation. *Bulletin of the World Health Organization*, **85**, 941–8.

Titmuss, R. (1997). *The gift relationship: from human blood to social policy*. London: LSE.

Trommelmans, L., et al. (2007). *Ethical issues in tissue engineering, European ethical-legal papers No. 7*. Leuven: Centre for Biomedical Ethics and Law. Available at: www.cbmer.be (accessed 7 September 2009).

Wadman, M. (2008). Blood transfusion for stem cell company? *Naturenews*. Available at: www.nature.com/news/2008/080822/full/news.2008.1060.html (accessed 7 September 2009).

Waldby, C. and Mitchell, R. (2006). *Tissue economies. Blood, organs and cell lines in late capitalism*. Durham, NC: Duke University Press.

Wendler, D. and Grady, C. (2008). What should research participants understand to understand they are participants in research? *Bioethics*, **22**, 203–8.

Wilkinson, S. (2005). Biomedical research and the commercial exploitation of human tissue. *Genomics, Society and Policy*, **1**, 27–40.

Williams, D.F. (2006). Tissue engineering: the multidisciplinary epitome of hope and despair. In: R. Paton and L. McNamara (eds), *Studies in multidisciplinarity*, Vol. 3, pp. 483–524. Amsterdam: Elsevier.

Chapter 16

The regulation of autologous stem cells in heart repair: comparing the UK and Germany[1]

Susanne Weber, Dana Wilson-Kovacs, and Christine Hauskeller

The expansion of biomedical research on novel medical applications using tissues and cells has been accompanied by increasing regulatory activity at the EU level over the last decade. As a result the EU tissue framework has been developed and medicinal product regulation extended.

Currently, there is little knowledge about the implementation of European regulations in existing national frameworks and how this shapes emerging biomedical fields of cell-based clinical research. This chapter provides a preliminary national comparison of the legal and institutional integration of the EU regulatory framework. It presents a case study of one of the few stem cell therapies that have moved towards clinical application—clinical trials using patients' own bone marrow stem cells, i.e. autologous cells, for heart repair. While European regulations have been established in order to harmonize the conditions for medical applications using tissues and cells across the continent, our analysis suggests that the ways in which they have been integrated in Germany and the UK reflect differing approaches to resolving uncertainty and managing risks in the development of novel cell therapies. These variations result in differences in local research practices and have implications for the intended harmonization process in Europe.

16.1 Analysing national regulatory contexts in clinical stem cell research

This chapter is the result of a comparative sociological study 'Stem Cell Research in Context' in which the impact of national and international governance practices on stem cell research in Germany and the UK was examined. Between 2007 and 2008 we carried out case studies on basic research with embryonic stem cells and clinical research with adult autologous stem cells in heart repair in both countries. Our findings are based on a textual analysis of European and national regulatory documents, ethnographic observations, and qualitative interviews with members of two research teams carrying out clinical trials with bone marrow stem cells in heart repair, and with representatives of regulatory authorities in the two countries.[2]

[1] The project 'Stem Cell Research in Context: A Comparative Study of the Dynamic Relationship between Science, Medicine and Society' (UK ESRC grant RES-329-25-0002) was conducted at the ESRC Centre for Genomics in Society (Egenis), University of Exeter. We thank the participants in this study for generously giving their time.

[2] Thirty-two qualitative interviews and six weeks of ethnographic observations were conducted in the two countries.

Comparative social studies of biomedical research and innovation have drawn fruitfully on a case study approach. By focusing on the in-depth analysis of salient cases, scholars have evaluated differences in the assessment and decision-making patterns adopted by regulators regarding technological risks in biomedical innovation (Daemmrich 2004; Abraham and Davis 2007; Faulkner 2009). Clinical research on autologous stem cells in heart repair constitutes a particularly instructive example of the implementation of EU tissue and medicinal product regulation in national jurisdictions,. It represents one of the few cases where stem cell research, as one of the areas currently considered most promising in the development of cellular therapies, has advanced to the clinical trial stage. Clinical trials employing bone marrow stem cells in heart repair are in progress in both Germany and the UK, thus allowing a cross-national comparison of the ongoing implementation of European regulations. Unlike embryonic stem cells, adult stem cells, especially autologous stem cells, have been considered to be ethically unproblematic (Hauskeller 2005; Faulkner et al. 2006). European regulatory practices that centre on the ethical risks of sourcing embryonic stem cells have been discussed elsewhere (Salter and Salter 2007), but the less ethically charged field of adult stem cell applications allows analysis of the national and transnational governance of the biological risks of novel medical applications using live cells.

Autologous stem cells fall between the biological risk categories established in European regulations. Patients' own bone marrow stem cells have been used for blood regeneration in transplantation and transfusion medicine in recent decades. The use of these cells in heart repair capitalizes on this established practice. This integration of routine procedures into a new application straddles the boundary between emerging regulatory categories of biological risk in more 'traditional' uses of tissues and cells in transfusion and transplantation medicine, and novel biotechnological ways of manipulating cells for the development of cell-based therapies (i.e. 'advanced therapies'). The ways in which these categories have been developed in Germany and the UK indicate the national rationales at work in the implementation of European tissues and cells regulations.

Below we present the background and general regulatory setting for clinical trials involving stem cells in heart repair. Then we describe the integration of European regulations on medicinal products and tissues and cells into German and UK regulatory processes and institutions, focusing on how biological risks are negotiated in each country. Finally, we discuss the implications of our findings for the harmonization of regulations involving novel cell therapies and identify further directions of research.

16.2 Background and regulatory setting of stem cell clinical trials in heart repair

Heart failure is one of the main causes of morbidity and mortality in Western societies, claiming significant health costs (Singh et al. 2009). Following increasing research into the regenerative capacities of stem cells in the human body (especially since the derivation of human embryonic stem cells in 1998), stem cells have been identified for both potential cardiac treatment and cardiac repair. A number of clinical trials involving patients with myocardial infarction or chronic heart failure in which patients' own bone marrow stem cells were used have been undertaken in Europe, East Asia, and the USA since the early 2000s (Lipinski et al. 2007; Charwat et al. 2008). The aim of these trials was to establish whether (and if so, under what conditions) stem cells might contribute to the regeneration of damaged tissue in the heart.

Several trials are currently ongoing in Germany (e.g. Strauer et al. 2002; Wollert et al. 2004; Assmus et al. 2006; Schächinger et al. 2006) and the UK (e.g. Galinanes et al. 2004; Yeo and Mathur 2009). They use similar testing procedures: bone marrow stem cells are extracted from the hip bone or flushed into the peripheral blood after drug-induced stimulation, and the stem cells are isolated from other cells and tissues (a routine laboratory procedure) and then injected into the damaged area of the patient's heart using different delivery methods. While results on the improvement of heart function have remained equivocal, the trials have been deemed safe (Bartunek et al. 2006).

The trials are designed within the strictures of evidence-based medicine. They follow increasing regulatory measures that centre on assuring the 'safety', 'quality', and 'efficacy' of the intervention. These parameters are defined in ethical, scientific, and technical guidelines and involve the management of risks through the documentation and monitoring of research according to principles of 'good practice'. They draw on standards of 'good clinical practice' (GCP) and 'good manufacturing practice' (GMP), for which guidelines have been developed at an international level since the 1990s.[3] Representing broad reference points for the regulation of medical applications under a risk-assessment paradigm, GCP and GMP are part of different regulatory and self-regulatory practices. In the EU special regulatory attention has been paid to the implementation of common ethical and technical provisions for the risk-controlled use of tissues and cells.

In practical terms, GCP guidelines entail the detailed specification of trial protocols and steps in the experimental procedure. The risks of this procedure are assessed on prior scientific evidence, the precise definition of a patient collective, and the outcome measures of the intervention (for different approaches to the assessment of risk, see Chapter 4). In order to ensure patients' autonomy regarding trial participation, informed consent procedures are required. Additionally, the trial itself has to be documented and a reporting system for adverse events devised to enable risk-based monitoring. The trial protocol, participant information sheets, and monitoring provisions must be presented in a formal application to an independent ethics committee, which represents a body of collective oversight during the trial and grants ethics approval.

In the preparation of cells for medical intervention, the protocol includes a description of their technical processing under risk-controlled conditions. Requirements for processing conditions, including documentation and monitoring procedures according to good practice principles, are subject to recent European regulations, where separate categories for cellular applications have been implemented. These categories distinguish between more or less 'risky' cells and involve measures of varying stringency with regard to the required degree of monitoring and control.

At this level, the regulatory procedures followed by German and UK research teams are similar: approval is needed from a local ethics committee and cell processing must be authorized by relevant authorities. However, the integration of European regulations into existing regulations and institutional structures for oversight has led to differing pathways in the practices of clinical teams and regulatory agencies. Below, we outline the emergent European framework and present the different routes taken by the two countries in the management of biological risk.

16.3 The EU framework: regulating novel cell therapies

Since the late 1990s the development of legislative measures at the European level has crystallized into two interconnected regulatory routes (Indech 2000). The first relates to public health concerns over the increasing globalization of exchange in human tissues. Legislative deliberations concerning patient safety resulted in the issue of the EU Tissue and Cells Directive in 2004 and accompanying technical directives in 2006.[4] The directives specify minimum ethical and

[3] These include standards issued by the International Conference on Harmonisation of Technical Requirements for Registration of Pharmaceuticals for Human Use (ICH) consisting of key pharmaceutical actors and EU, Japanese, and US regulatory agencies.

[4] Directives 2004/23/EC OJ L102/48; 2006/17/EC OJ L38/40; 2006/86/EC OJ L294/32. Organs are excluded from the framework.

technical requirements regarding consent procedures in the donation of tissues, and standards for their procurement, testing, preservation, storage, and distribution to patients in order to prevent disease transmission and contamination. The criterion of traceability of tissues and cells from donor to recipient and the licensing of establishments handling them by regulatory authorities have been implemented to enforce standards of risk control (Kent et al. 2006a).

The second regulatory route centres on the expansion of innovations in the production of cellular therapies and tissue engineering (see Chapter 15). These may entail a combination of live tissue currently classified as 'biological', synthetic structures classified as 'medical devices', and drugs classified as 'pharmaceuticals', and expand beyond the scope of existing regulatory frameworks for medicinal products, medical devices, or biologicals. Here, efforts have focused on resolving regulatory inconsistencies in the definition of such products and establishing appropriate controls for product safety to facilitate uniform marketization across Europe. Rather than implementing a separate framework for tissue-engineered products, provisions for all cell-based therapies were integrated into existing EU medicinal products regulation. Consequently, cell-based therapies have been aligned more closely with pharmaceutical products under the auspices of the European Medicines Agency (EMEA) as the established European agency of oversight. This change has been interpreted as ensuring rigorous regulation of future cell therapeutics (Faulkner 2009).[5]

The EU tissue and cells directive and the elaboration of medicinal product regulation around cell-based therapies have resulted in an intersecting two-tier framework. Cells and tissues fall under medicinal product regulation when they are classified as 'somatic cell therapy', 'gene therapy', or 'tissue-engineered products'. The former two categories were established in 2001, and the definition of a tissue-engineered product was introduced in the Advanced Therapies Regulation which came into force in December 2008 and seeks to establish clear demarcation and provisions for all cell-based products under the new heading of 'advanced therapies'.[6]

If tissues and cells are categorized as 'advanced therapies', the EU tissue and cells framework governs only their donation, procurement, and initial testing. Their further processing, storage, and distribution are regulated under medicinal product law. If tissues and cells are not classified as falling within one of these three categories, but are used in human application, they fall exclusively under tissue and cells regulation, covering the entire process from donation to distribution (see also Chapter 15).

The rationale behind this distinction has been to maintain a separation between the more 'traditional' use of tissues and cells in a hospital environment, centred around transplantation practices (e.g. bone marrow) or the transplantation of body parts (e.g. corneas and heart valves), and newer techniques of biotechnological manipulation and engineering which involve cells and tissues as living material to 'manufacture' medical applications (Kent et al. 2006b). The distinction is based on assumptions about the biological risks entailed by these uses of cells and the subsequent needs for control and monitoring. Risks are classified according to whether the cells or tissues are 'worked on' and whether their 'nature' could have changed in this process.

---

[5] A third regulatory concern relates to the removal, storage, and use of human tissues for biomedical research purposes (not involving human application), which is addressed in other chapters in this volume regarding the regulations for the protection of individual autonomy and privacy and the commercialization of human body parts in the use of human tissues for research. For UK and German contexts, see especially Chapters 5, 8, and 14.

[6] Directives 2001/83/EC OJ L311/67; 2003/63/EC OJ L159/46. Regulations (EC) No 726/2004 OJ L136/1; (EC) No 1394/2007 OJ L324/121.

Accordingly, technologies which change the genetic make-up of cells, enhance their growth in culture by adding biomolecules, or combine them with synthetic scaffolds are seen to carry more potential health risks. In contrast, the technical preparation of cells, which is common in transplantation medicine, is understood to be less risky and not to alter the nature of the cells. Although subject to the conditions specified in the EU tissue framework, the technical requirements for risk control are not as extensive as in the case of GMP provisions for 'advanced therapies'.

Concomitantly, the medicinal product regulations set out the notions of 'industrial process' and 'substantial manipulation' to distinguish riskier 'advanced therapies' from less risky medical procedures involving tissues and cells. Furthermore, if cells or tissues are to be used for a purpose in the patient's body that is different from what they usually do, they fall into the class of 'advanced therapies'. However, the application of these criteria is open to interpretation across Europe. The regulators we interviewed talked about the uncertainty in classifying cellular applications in their daily work.

As a result of the two-tier structure of clinical trial regulations for cell therapies, GCP requirements established in EU medicinal products regulation in 2001 only apply to trials involving cells and tissues classified as 'advanced therapies' and do not pertain to those with less 'risky' tissues and cells.[7]

16.4 **The UK framework: assemblage**

In the UK the EU regulation structure has been implemented as two regulatory frameworks for medicinal products and tissues and for cells in human application. Two agencies have been designated as competent authorities.[8] Since 2004 the regulation of medicinal products legislation has been based on the Medicines Act 1968.[9] The Medicines and Healthcare Products Regulatory Agency (MHRA), established in 2003 by merging two previous bodies (the Medical Devices Agency and the Medicines Control Agency), regulates medicinal products and the use of cells for 'advanced therapies'. EU tissue regulations were integrated into the framework of the Human Tissue Act 2004 to regulate the removal, storage, and use of human tissue. The Human Tissue Authority (HTA) was set up as statutory body under the Act in 2005 and extended its remit in 2007 to become the competent authority under EU tissue regulation for less 'risky' tissue and cells practices (see Chapter 8).[10]

Depending on the classification of tissues and cells, the HTA only oversees their donation, procurement, and initial testing; the MHRA oversees further steps in the processing and application when it comes to 'advanced therapies' clinical trials. Such trials are overseen by a specially instituted ethics committee for gene and stem cell therapies, the Gene Technology Advisory Committee (GTAC). Trials with tissues and cells which are not classified as 'advanced therapies' fall within the remit of the HTA. Following the logic of the EU framework, when trials are classified as not falling within medicinal product regulation, they also remain outside the purview of EU GCP regulations. The gap in the oversight of clinical research practices involving tissues and cells is then filled by the Research Governance Framework of Health and Social Care by the Department

[7] GCP provisions are specified in 2001/20/EC OJ L121/34 and 2005/28/EC OJ L91/13.

[8] The Human Fertilization and Embryology Authority (HFEA) oversees the use of and research on germ cells in reproductive medicine.

[9] Medicines for Human Use (Clinical Trials) Regulations 2004 (SI 2004 No 1031); Medicines for Human Use (Clinical Trials) (No.2) Regulations 2006 (SI 2006 No 2984).

[10] Human Tissue Act 2004 (*c.* 30); Human Tissue (Quality and Safety for Human Application) Regulations 2007 (SI 2007 No 1523). Additionally, the HTA has issued nine Directions since 2006.

of Health, which is also oriented towards GCP principles (Department of Health 2005), and good practice guidance issued through the National Health Service and other bodies such as the Medical Research Council.

Thus the conduct of clinical trials employing stem cells in the UK has become set within two variable processes that are centred on a graded approach to overseeing the risks posed by cells in medical applications. This approach involves an assemblage of authorities and mechanisms which take over from each other in setting technical conditions for cell processing and carrying out trials according to GCP principles. The ensuing interfaces between authorities and processes of oversight require negotiations between different actors regarding the classificatory status of tissues and cells and the appropriate regulatory mechanisms.

In our example, clinical trials with autologous bone marrow stem cells in heart repair are considered 'less risky'. The trials started before the HTA assumed its remit under EU tissue law, and, in accordance with the regulations at the time, the cells were processed in a haematological laboratory accredited as meeting GMP requirements under the voluntary Code for Tissue Banks (Department of Health 2001). Subsequently, this laboratory has been licensed by the HTA. Although the trials were approved by an NHS ethics committee under the Department of Health Research Governance Framework, technically they were not covered by EU GCP provisions. The incremental development of regulatory control illustrates the emergent nature of UK governance processes and highlights the logic of assembling different regulatory mechanisms for the oversight of medical applications using tissues and cells as part of the implementation of EU regulations.

16.5 **The German framework: streamlining**

In Germany, provisions regarding both 'advanced therapies' and the use of cells and tissues have been integrated into a single legislative instrument, the German Medicines Act 1976. Tissues and cells (except germ cells, blood and organs) count as medicinal products (*Arzneimittel*) as a matter of principle. In regulatory terms this means that EU GCP provisions apply to all clinical trials (Schriever et al. 2009). Moreover, in implementing EU tissue regulations, the notion of 'tissue preparation' (*Gewebezubereitung*) was introduced into the Medicines Act in 2007, with the German Tissue Law, to distinguish between 'advanced therapies' and less risky cell practices.[11] Consequently, since 2004–2005, when EU medicinal products regulation was subsumed into the Medicines Act and set out in GCP regulations, all clinical trials using tissues or cells in Germany (not just those involving 'substantial manipulation' or an 'industrial process', as in the UK) have had to be authorized by the national competent authority (the Paul-Ehrlich-Institute) in conjunction with a federally approved ethics committee. Since 2004, in the processing of cells, the approval procedure for the production of medicinal products has been extended to include riskier 'somatic cell therapies' and 'gene therapies' (the category of tissue-engineered product has only recently been introduced into European regulation) and the less risky 'tissue preparations' (Pruss 2008). The preparation of cells has to be approved by the relevant federal authority of the German state in which a trial is located. This approval is issued in concordance with the Paul-Ehrlich-Institute. Thus the Paul-Ehrlich-Institute is the main authority in the regulatory process, which has continuously extended its remit since 1972.

Thus clinical trials using bone marrow stem cells in heart repair are not subject to the same assemblage of variable regulatory mechanisms and authorities around the principles of GCP and

[11] 12th and 14th revisions of the Medicines Act in BGBl I p. 2031, 5.8.2004 and BGBl I p. 2570, 5.9.2005; Tissue Law in BGBl. I p. 1574, 27.7.2007. The Act's provisions are elaborated in specific implementation regulations.

technical requirements for cell processing as they are in the UK. The integration of EU regulations into the German framework has also entailed the introduction of new regulatory processes and been accompanied by uncertainty and negotiations among federal authorities, the Paul-Ehrlich-Institute, cell-processing units, and researchers. This integration has built on pre-existing institutional structures rather than requiring the establishment of new regulatory agencies. Regulated under the auspices of one main authority and one main legal instrument, clinical trials are subject to a more streamlined oversight procedure. The integration of tissue and medicinal product regulation into the Medicines Act means that both 'more risky' and 'less risky' cellular therapies exist on a continuum and are set within an institutional and legal structure of oversight which seeks to contain biological risks in a more uniform regulatory apparatus that includes pharmaceutical drugs and human tissues and cells. In our case, , according to our interviews with German regulators, clinical trials using bone marrow cells in heart repair, currently rated as 'less risky' in the UK, might be classified in future as 'somatic cell therapy', i.e. 'more risky' (see also Schilling-Leiss et al. 2008).

16.6 **Conclusion**

Our analysis suggests that Germany and the UK have taken different routes in regulating biomedical developments in novel cell therapies. Germany has adopted a streamlined approach in which a single regulatory body oversees the integrity of research as well as tissue practices. Its legislative framework aims to be comprehensive by classifying all tissues and cells in human application as medicinal products. In contrast, the UK's regulatory structure separates tissues and cells on the basis of their assumed biological 'riskiness'. This is complemented by an assemblage of institutional mechanisms set up to ensure the ethical and risk-controlled conduct of research—as evidenced by the MHRA, the HTA, the GTAC, the Department of Health Research Governance Framework, and non-statutory good practice guidance. If we had considered embryonic stem cells, additional bodies would be involved, namely the Human Fertilization and Embryology Authority (HFEA) and the Stem Cell Bank (Hauskeller 2004).

These differing approaches affect the ways in which the clinical research teams we interviewed relate to regulation. While German team members talked about regulation in terms of the formalization of procedures, the UK team stressed the emerging nature of regulatory practices and the process of bringing together different actors and agencies in the regulation of their trials. These interviews illustrate different phases in the routinization of clinical research practices, which relate to the fact that clinical trials have been conducted for a longer period of time in Germany than in the UK. Most significantly, they demonstrate how regulatory environments shape views on regulation and its implementation in everyday research. We have shown elsewhere (Wilson-Kovacs et al. 2010) how clinical research teams in both countries strategically frame regulatory practices in different contexts to justify their trials as a legitimate research endeavour in an emerging regenerative medicine field.

In order to understand whether the more streamlined approach adopted in Germany and the process of assemblage in the UK translate into substantive differences in the development of cell-based biomedical applications in Europe, a close analysis of the classificatory practices surrounding borderline cases such as autologous bone marrow stem cells in heart repair is needed. Our data suggest that the classification of these clinical trials might take different routes in the two countries. Whereas in the UK the processing of the cells is currently classified as less risky medical practice, the statements of German regulators indicate a move towards placing them in the riskier category of 'somatic cell therapy'.

The local classification of boundary cases according to different categories provided by European regulations has implications for the harmonization of tissue practices in terms of requirements for cell processing and clinical practice across Europe, and the developmental pathway of cell-based therapies. If the use of autologous stem cells in heart repair were to move beyond the clinical trial stage into medical practice under EU tissue law in the UK, no further authorization would be needed at the European level. In contrast, following the classification as 'somatic cell therapy' in Germany, this use would be subject to marketing authorization at the European level through the EMEA centralized procedure for 'advanced therapies'. These possible divergences show that while the aim of Europe-wide regulations for medicinal products and tissues and cells has been to facilitate uniform conditions, their implementation will require detailed engagement with national and local situations.

In this respect the Committee for Advanced Therapies (CAT), established at the EMEA in late 2008 as a result of the Advanced Therapies Regulation, is of special interest. The remit of this multidisciplinary body, which consists of experts from national authorities, is to issue guidance, draw up technical guidelines, and provide scientific dossiers for the assessment of marketing authorizations of cell-based therapies in collaboration with the Committee for Medicinal Products for Human Use (CHMP) to the European Commission. In order to account for the difficulties in the categorization of cell-based therapies and to achieve the harmonization of tissue practices, the Committee is working towards a system of procedural advice on the classification of 'advanced therapies' for developers of cell-based therapies.[12]

In addition to the analysis of the development of regulatory practices and cell-based therapies within national contexts, understanding the practices of CAT will provide insight into how views on the classification of cell-based therapies are framed, elaborated, and fed back into national practices. An evaluation of the interchange between local, national, and European governance practices may highlight the precise points at which differences in classification and regulation occur and how these differences are negotiated among actors. Such an analysis will also allow a broader assessment of Europe-wide harmonization of regulations in biomedical research and clinical applications, and of their integration into and function across national contexts.

References

Abraham, J. and Davis, C. (2007). Deficits, expectations and paradigms in British and American drug safety assessments: prising open the black box of regulatory science. *Science, Technology and Human Values*, **32**, 399–431.

Assmus, B., Honold, J., Schächinger, V., et al. (2006). Transcoronary transplantation of progenitor cells after myocardial infarction. *New England Journal of Medicine*, **355**, 1222–32.

Bartunek, J., Dimmeler, S., Drexler, H., et al. (2006). The consensus of the task force of the European Society of Cardiology concerning the clinical investigation of the use of autologous adult stem cells for repair of the heart: ESC Report. *European Heart Journal*, **27**, 1338–40.

Charwat, S., Gyöngyösi, M., Lang, I., et al. (2008). Role of adult bone marrow stem cells in the repair of ischemic myocardium: current state of the art. *Experimental Hematology*, **36**, 672–80.

Daemmrich, A. (2004). *Pharmacopolitics: drug regulation in the United States and Germany*. Chapel Hill, NC: University of North Carolina Press.

Department of Health. (2005). *Research governance framework for health and social care*. London: Department of Health.

Department of Health. (2001). *A code of practice for tissue banks providing tissues of human origin for therapeutic purposes*. London: Department of Health.

---

[12] EMEA/99623/2009.

Faulkner, A. (2009). *Medical technology into healthcare and society: a sociology of devices, innovation, and governance*. Hampshire: Palgrave Macmillan.

Faulkner, A., Kent, G., Geesink, I., and FitzPatrick, D. (2006). Purity and the dangers of regenerative medicine: regulatory innovation of human tissue-engineered technology. *Social Science and Medicine*, **63**, 2277–88.

Galinanes, M., Loubani, M., Davies, J., Chin, D., Pasi, J., and Bell, P.R. (2004). Autotransplantation of unmanipulated bone marrow into scarred myocardium is safe and enhances cardiac function in humans. *Cell Transplantantation*, **13**, 7–13.

Hauskeller, C. (2004). How traditions of ethical reasoning and institutional processes shape stem cell research in Britain. *Journal of Medicine and Philosophy* **29**, 509–32.

Hauskeller, C. (2005). Science in touch: functions of biomedical terminology. *Biology and Philosophy*, **20**, 815–35.

Indech, B. (2000). The international harmonisation of human tissue regulation. *Food and Drug Law Journal*, **55**, 343–72.

Kent, J., Faulkner, A., Geesink, I., and FitzPatrick, D. (2006a). Towards governance of human tissue-engineered technologies in Europe. *Technological Forecasting and Social Change*, **73**, 41–60.

Kent, J., Faulkner, A., Geesink, I., and FitzPatrick, D. (2006b). Culturing cells, reproducing and regulating the self. *Body and Society*, **12**, 1–23.

Lipinski, M., Biondi-Zoccai, G.G., Abbate, A., et al. (2007). Impact of intracoronary cell therapy on left ventricular function in the setting of acute myocardial infarction. *Journal of the American College of Cardiology*, **50**, 1761–7.

Pruss, A. (ed.). (2008). Das deutsche Gewebegesetz—Anforderungen und Chancen für Gewebebanken und klinische Anwender (The German tissue law—demands and chances for tissue banks and clinical operators). *Transfusion Medicine and Hemotherapy*, **35**, 402–82

Salter, B., and Salter, C. (2007). Bioethics and the global moral economy. *Science, Technology and Human Values*, **32**, 554–81.

Schächinger, V., Erbs, S., Elsässer, A., et al. (2006). Intracoronary bone marrow—derived progenitor cells in acute myocardial infarction. *New England Journal of Medicine*, **355**, 1210–21.

Schilling-Leiss, D., Godehardt, A.W., Scherer, J., et al. (2008). Genehmigungsverfahren für klassische Gewebezubereitungen gemäß § 21a Arzneimittelgesetz (AMG) (Approval procedures for classical preparations of tissue). *Transfusion Medicine and Hemotherapy*, **35**, 453–62.

Schriever, J., Schwarz, G., Steffen, C., and Krafft, H. (2009). Das Genehmigungsverfahren klinischer Prüfungen von Arzneimitteln bei den Bundesoberbehörden (The approval procedure of clinical trials for medicinal products). *Bundesgesundheitsblatt*, **52**, 377–86.

Singh, S., Arora, R., Handa, K., et al. (2009). Stem cells improve left ventricular function in acute myocardial infarction. *Clinical Cardiology*, **32**, 176–80.

Strauer, B.E., Brehm, M., Zeus, T., et al. (2002). Repair of infarcted myocardium by autologous intracoronary mononuclear bone marrow cell transplantation in humans. *Circulation*, **106**, 1913–8.

Wilson-Kovacs, M., Weber, S., and Hauskeller, C. (2010). Stem cells clinical trials for cardiac repair: regulation as practical accomplishment. *Sociology of Health and Illness*, **32**, 89–105.

Wollert, K.C., Meyer, G.P., Lotz, J., et al. (2004). Intracoronary autologous bone-marrow cell transfer after myocardial infarction: The BOOST Randomized Controlled Clinical Trial. *Lancet*, **364**, 141–8.

Yeo, C. and Mathur, A. (2009). Autologous bone marrow-derived stem cells for ischemic heart failure: REGENERATE-IHD Trial. *Regenerative Medicine*, **4**, 119–27.

Chapter 17

Discovering informed consent: a case study on the practices of informed consent to tissue donation in Austria[1]

Milena D. Bister

17.1 Introduction

With the rise of biomedical technosciences and the completion of the Human Genome Project, tissue collections worldwide have become of the utmost scientific and economic interest. Recently, questions relating to donors' informed consent have become of great interest particularly with respect to the procurement of new tissue samples for biobanking. Most of the public and bioethical debates are centred on the need to safeguarding donors' rights on autonomy (for an alternative approach, see Chapter 2). Whereas article 22 of the Council of Europe's Convention on Human Rights and Biomedicine clearly states that supplementary informed consent must be obtained for utilization of tissue leftovers for purposes other than those originally intended, international efforts to harmonize biobanks do not insist upon informed consent (e.g. OECD 2007; Yuille et al. 2007). Currently, Austrian law also allows consent of the individual donor to be bypassed in such cases. At university hospitals in particular, a hypothetical interpretation of the contract of treatment between patients and doctors implies an 'implicit agreement' of the silent patient on the legally specified institutional objectives of teaching and researching (Kopetzki 2004).[2] In contrast with the legal perspective, research ethics committees and medical researchers increasingly have recourse to the well-established tool of informed consent in such matters (for example, see Chapter 14 for a suggested formalized consent process). It seems that informed consent has been advanced as the missing ethical solution to the many challenges at the interface between clinical duty and research interest (Hoeyer 2008).

The ways in which informed consent to the donation of tissue leftovers is framed—on the one hand by law and on the other hand by clinical practice—show that there is more to consent than issues of litigation or donors' individual autonomy. As already noted, discussions about autonomy have been central to standard bioethical work on informed consent (Faden and Beauchamp

[1] I gratefully acknowledge the material and intellectual contribution of my former colleagues Ulrike Felt, Michael Strassnig, and Ursula Wagner to the work that has led to this chapter. Particular thanks go to Ulrike Felt for supervising my PhD and to Klaus Hoeyer for making helpful comments on an earlier draft of this chapter. I am also grateful to all patients and professionals who participated in this study for sharing their views. I thank the Austrian Federal Ministry for Education, Science, and Culture (TRAFO Programme) for funding the informed consent project and the University of Vienna for its support of my PhD through a personal research grant (F78-S) in 2007.

[2] Several exceptions to informed consent are also included in the Swiss draft bill of the act on research involving humans (see Chapter 10).

1986; Berg et al. 2001). They also constitute the explicit conceptualization of consent in transnational biobank projects (OECD 2007). However, the literature primarily generated by qualitative social scientists has been concerned with debating the *circumstances* surrounding participants' consent to tissue donation (Corrigan 2003; Busby 2004; Hoeyer and Lynöe 2006; Dixon-Woods et al. 2007; Felt et al. 2009). It has also been suggested that the focus on biobanks and informed consent as *ethical* matters has amounted to facilitation rather than regulation of research (Black 1998; Hoeyer and Tutton 2005).

In this chapter I will examine one aspect of tissue donation in Austria, namely the donation of tissue leftovers from plastic surgery. As a result of empirical analysis, my aim is to explore issues of informed consent by highlighting how consent interactions between the researcher and the prospective donors work as purposive procedures. From that perspective we can see how informed consent relates not only to simply respecting donors' rights of autonomy, but also to a variety of other objectives in. In this way I aim to contribute to further discussion of the dynamics of informed consent, particularly regarding the efforts of harmonizing legal and ethical questions with respect to research using human bodily substances.

17.2 **Methods**

The results presented in this chapter are based on an empirical research project on informed consent to tissue donation in Austria, which was conducted between 2005 and 2007 in cooperation with one of Austria's largest university hospitals[3]. In collaboration with a medical researcher, who conducted informed consent, and her supervisor, a clinical pathologist, we studied the situation in which patients were asked for consent to a basic research project where tissue removed during plastic surgery would be used as research material for analysing the molecular mechanisms of diabetic microangiopathy. Informed consent was not obligatory but was recommended by the hospital's ethics committee. Patients had been admitted to the hospital for breast size reduction, breast reconfiguration after cancer treatment, or abdominoplasty, and their tissue would provide the researcher with necessary comparative data as it was obtained from non-diabetic individuals.

The outcomes presented here are mainly based on observations of 24 informed consent conversations between the researcher and the patients as well as subsequent semi-structured qualitative interviews with 19 patient-donors.[4] In addition, results are based on expert interviews with the medical researcher, the clinical pathologist, and a member of the research ethics committee, and on a group discussion with experts from various fields (patient groups, law, ethics, sociology, medicine). It is noteworthy that we ourselves had to obtain the patient's written consent before starting the interview. Therefore we found ourselves in a similar situation to the one we observed between the medical researcher and the patients. Observation protocols have been analysed following the coding paradigm of grounded theory (Strauss and Corbin 1990). Interviews were tape-recorded and transcripts were interpreted according to qualitative content analysis via inductive category development (Mayring 1997).

[3] The project 'Informed consent: space of negotiation between biomedicine and society' was conducted at the Department of Social Studies of Science at the University of Vienna by Ulrike Felt, Michael Strassnig, Ursula Wagner, and Milena Bister. It received funding by the Austrian Federal Ministry for Education, Science and Culture within the TRAFO (Transdisciplinary Research) programme.

[4] Three of the 24 patients refused to give an interview. Two interviews were not been tape-recorded because of technical failure and were excluded from later analysis.

Tissue donation in this study was not destined for contribution to a biobank project. Nonetheless, the situational context of informed consent explored here will be encountered elsewhere, especially when the conditions under which patients and researchers are involved are similar to those formulated below. This is likely for consent procedures regarding tissue banking.

In the following I will delineate how the patients and the medical researcher participating in this case study approach informed consent. Then I will identify how those observations and interpretations promote the perception of the consent process as a purposive procedure and how that might invite rethinking the concepts and dynamics of consent in biomedical research activities using leftover human tissue.

17.3 **Readiness of address: informed consent before undergoing surgery**

During both discussion of informed consent to tissue donation and the subsequent interviews the forthcoming surgery is at the centre of the patients' concerns. Their narratives mainly describe the medical necessity of the intervention because of physical pain (e.g. back-ache) or impediments to a sporting/healthy lifestyle or other social interactions. In their accounts they describe, and simultaneously constitute, corporeal and social dimensions related to their hospital stay which draw upon the prevalent cultural practice of legitimating plastic surgery (Gimlin 2007). The potential double function of solving both medical needs and aesthetic demands, which is characteristic of breast reduction surgery and abdominoplasty in general, disappears in the patients' accounts, leaving almost exclusively descriptions of medical indications. However, previous uncertainties regarding the medical indication of the operation are present in some interviews, as shown in the extracts from interviews quoted below.

> Q: What were you talking about [with the surgeon]?
> A: Everything, starting with examination, you know, when he ascertained that it [the intervention] is really necessary, you know, and not just my imagination [*smile*] One is wondering, indeed. That's what I wanted to have explained by him, and by other doctors as well [*smile*]
> Female patient, age 31–45, breast size reduction

Concerns regarding the legitimacy of the intervention decrease on the day before the operation. Now, the decision to undergo surgery will be endorsed step by step in formal acts, such as examinations and enquiries. Therefore the day before surgery is already part of its performance. In the words of Bourdieu, how we experience the present depends as much on our expectations of the future as on what we have experienced in the past (Bourdieu 2001). Hence, the surgery, though still to come, already constitutes the present. It is in the interest of patients to ensure that everything proceeds smoothly. They are careful to contribute as best they can to a good ambience as well as to facilitate the processing of their medical case by collaboration. With that aim patients are available to the hospital staff at virtually any time. They are prepared to fill in and sign forms, to collaborate in examinations, and to exchange detailed information. This condition of readiness to collaborate fully in the completion of such tasks, which I will call 'readiness of address' (*Ansprechbereitschaft*), is fundamental if we wish to understand more of the practice of informed consent from a patient's perspective.

During readiness of address, the patient's prime activity is waiting for the next interaction. 'Waiting' in this context does not mean being inactive, but rather carrying out activities that can be stopped at any time. Some events might be expected, such as giving informed consent to surgery, while others might come as a surprise in terms of the purposes they serve. Since the informed consent procedure concerning tissue donation takes place on the day before operation, i.e. in the context of forthcoming surgery, its occurs when patients are in the state of readiness of address. This is extremely important as it has a strong influence on patients' assessment of informed consent and subsequently on their decision-making.

Although the request for informed consent to tissue donation surprised most of the patients in our study, they did not find it difficult to understand its requirements, because informed consent is similar to other interactions with medical staff members. In the following extract the patient expresses how she experiences informed consent to tissue donation as a normal occurrence in the clinic:

> There is a lot occurring right before surgery, well, that's, whether there is one more or less, it is always the most hectic day...that one is coming and this one is coming, then still this and still that and something else. It does not matter...whether or not one more person. It is no problem for me.
> Female patient, age 46–60, breast reconfiguration (P02:379)

A request for consent to tissue donation is a novel event for patients, but it is similar to their previous experiences within the health-care system. Consequently, informed consent does not stop the patients' readiness of address, but confirms that specific condition. From a patient's perspective informed consent to tissue donation falls into line with other medical and nursing activities which they accept so that their surgery can be performed. Although they are told by the researcher that this particular informed consent is not part of the overall preparations for the operation itself, but concerns the tissue left over from surgery, the manner in which the researcher asks for informed consent resembles other acts which do concern the medical intervention. This does not imply that patients consent *without understanding* that their surgery will not be affected. Exactly the opposite is the case: patients know that the donation of tissue will not change the surgical procedure. As shown in the following extract, patients can distinguish between procedures that involve physical risk and those that do not.

> What happens with the tissue then doesn't matter to me, because it doesn't belong to me. But at the moment that the doctor operates on me it is my body...it is a big difference. Because the tissue that is removed is dead with respect to me. But that still in me is not dead, you know.
> Female patient, age 26–30, breast size reduction (P16:175)

Patients do not confuse tissue donation with the surgical intervention. The similarity of acts is much more important to the process of decision-making itself. In the context of their readiness of address patients assume on principle that they will collaborate and consent. Hence, in practice, the option *not to consent* remains in the background (Felt et al. 2009). In our study all patients agreed to the use of left-over tissue in research. Surprisingly, the actual content of the research project is irrelevant to patient consent. Patients evaluate informed consent by their own criteria (Felt et al. 2009). Their central concerns are not reasons to support the specific research methods and goals in question, but rather the absence of personal motives that justify a refusal of consent.

In addition, in the state of readiness of address, patients experience informed consent not primarily as an intrinsic moment of decision, but as a kind of *request* (Dixon-Woods et al. 2006). According to their own reasoning, they can agree to the researcher's request without hesitation. One patient phrased it as follows.

> Well, there is a difference in that case of course. It's surely something that one does not need to do, but all stuff, what has been signed so far, it essentially needs to be signed. You are informed, I admit, but in principle you need to sign. In this case you are not obliged. This is just a request.
> Male patient, age 26–30, abdominoplasty (P19:119)

The perception of informed consent as a request certainly affects the patients' decision-making during the procedure. Two socially distinct types of requests prevail in the interviews: first, the request to help the researcher's work with a personal contribution; secondly, the request to support the hospital's research activities in general (Bister 2009). Whereas the first type explains consent in the logic of an interpersonal relationship between researcher and patient, the second draws on a broader relation between the patient and the hospital as a part of the public health system (Felt et al. 2009). The practice of informed consent enables the patients in our study to comply with the request of the researcher and the hospital to donate their tissue, and they do so without hesitation provided that their personal interests are not impaired.

17.4 **Managing uncertainties: informed consent before performing research**

What about the researcher's perspective on informed consent? To her, informed consent to tissue donation represents a repeated act. Once she has obtained overall approval of the research project from the hospital ethics committee, she has to start a cascade of different steps that need to be well coordinated when additional research material is required. Cooperation with the nursing staff is extremely important. Not only does it save the researcher time, but their collaboration is necessary to secure that the signed consent form will actually result in deposition of the tissue at the surgery. If theatre nurses fail to notice the patient's informed consent sheet, the researcher will not receive any tissue. Therefore it is essential to ensure cooperation from the nursing staff. Sometimes, despite assistance from the nurses, the researcher still does not obtain any research material. This occurs when surgeons ignore the consent form and decide, based on their own convictions, that the tissue will not be donated to research. In other cases, even if the researcher receives the tissue as expected, it is not always suitable as research material. For instance, the tissue may have been excised in pieces that are too small for preparation. There is nothing that the researcher can do about this, except wait for the next appropriate patient and start the procedure again.

Therefore the uncertainties regarding tissue procurement do *not* reside in informed consent. As previously noted, during our collaboration with the researcher all patients consented, as was also the case throughout her medical research project. Although the researcher had no control over uncertainties that occur in the operating theatre, she had some influence on the patients' decision-making process regarding informed consent. By her mere presence she is able to react to any insecurity, thereby participating in the decision. An interview with a member of the research ethics committee, who is also a hospital physician, revealed that a high consent rate is not specific to the particular informed consent procedure we have studied. It is customary in this hospital for patients to donate bodily substances to research. She explains it as follows:

> Generally speaking we are in the fortunate position that there is still a foundation of trust...Well, we are conducting genetic examinations on a large scale. And we have...since that particular ordinance... has been enacted—that's, I believe, for about two years now—that patients who don't want their material to be used, can statee this and we [the ethics committee] then need to be informed. And over the whole period, since that ordinance became effective, I have received only one message that a patient did not want the material be used in research.
> Member of the ethics committee, physician (KI–13:95)

Therefore high consent rates are not unusual. The uncertainty as to whether patients will consent to tissue donation, which in principle exists until the patient makes that particular decision, disappears in the realities of the hospital. The researcher's presence during the patient's decision-making process results in conflicts of interests which are well known in doctor–patient interactions in general (Berg et al. 2001). The researcher's supervisor, himself a pathologist at the hospital, pointed out that in his opinion bias is intrinsic to any medical informed consent:

> [I]t is precisely a question of presentation...Of course I will describe chemotherapy in a way that he will take it...always there will be some decision preferred ...I also wish them to take this, you know... That is, I think that empowerment is anyway, will probably never be based upon objective grounds, but always be very biased concerning the opinion of traditional medicine or medicine at this point of time...
>
> Pathologist, supervisor of the researcher, expert group discussion (EG-2:1084)

This conflict is also present during informed consent to tissue donation. The researcher's interest in obtaining tissue comes across with the imperative of objective information. The following extract illustrates how the researcher resolves that conflict:

> Because he [the patient] just wants to know in fact what he needs to know, what he needs to do, sort of, what the demands are. After all I come to him with a question, with a request...if the results were that everybody consented because he somehow wanted to help research then I think that's only good. I think it is good, it's good anyway. Because it would be a result that somehow strengthens you as a researcher, because of course you don't want to do any harm, you simply want to make them understand that they can help you if they sign now and donate this, something like that.
>
> Researcher (IR:335)

Hence the researcher's understanding of the matter mirrors a notion of *requesting* personal support while performing informed consent. In the same vein, the member of the ethics committee quoted earlier refers to a practice of request when reflecting upon the political difficulties inherennt in introducing procedures for global consent:

> And since this [a global informed consent procedure with restrictions to specific areas of research] is obviously somehow still much more difficult to administer it now occurs essentially in a way...that everybody shapes one's study in so far as he says: 'I would like to investigate *this*, please give me your consent. I would like to investigate *that*, please give me your consent. I would like to investigate *this*, please give me your consent'.
>
> Member of the ethics committee, physician (KI-13:25)

For these reasons we can say that the procedure of informed consent to tissue donation, which in practice is put as a request, is not an exception to general hospital routine. Although it serves the specific purposes of the institution, informed consent to tissue donation could be viewed as an example of overlapping agendas in the clinic (e.g. treatment, research, patient care) rather than as a case of a unique encounter between the researcher and the donor.

17.5 **Discussion**

If we wish to advance the arguments above, informed consent might be seen as beneficial for different purposes depending on whether one is awaiting surgery or planning research. By assessing informed consent as, for example, an opportunity to contribute to the medical enterprise, most patients welcome informed consent on their path to surgery as a way of positioning themselves 'positively', i.e. cooperatively, with respect to the medical system.[5] Some patients enjoy the

---

[5] Patients have experienced and 'utilized' our social science interviews in similar ways. Like tissue donation, our interviews have not been part of the patients' prime interest of receiving care. In the same way as the researcher introduced the realm of medical research to the patients, we have introduced them to the realm of social science.

researcher's personal attention and efforts to tell them that research will be undertaken with their leftover skin tissue (Felt et al. 2009). To the researcher, informed consent is the key to obtaining access to tissue as research material in an institutionally legitimated way. Furthermore, the informed consent procedure facilitates publication in major scientific journals, which is indispensable for a successful career.

Considering all this, it seems fruitful to re-conceptualize informed consent as a purposive institutional procedure to which researchers as much as patients devote themselves (Hoeyer 2009). This will allow the experiences of both researchers and patients to be included in understanding the practices and the limits of informed consent. From that perspective it might become possible to assess how procedures of informed consent in general have been affirmed in such diverse medical arenas as therapy, research, clinical trials, and biobanks. Taking into account the interests of those *in touch* with informed consent might facilitate an understanding of its extensive use currently. The concept of 'boundary objects' (Star and Griesemer 1989) may reveal how diverse actors manage to collaborate across their interests. In their study of the establishment of a natural history research museum in Berkeley, Star and Griesemer argue that establishing boundary objects is a common practice applied by, for instance, amateurs and professionals to translate between their social worlds during cooperation. For example, the species of mammals became one of the boundary objects, as both groups had a common interest in including them in the museum collections. Although the species had different meanings in the two communities, they were familiar to all and therefore provided appropriate tools for communication. Extending the concept from material objects to techniques as well as stories and ideas, the authors summarize that the special character of boundary objects allows them to be 'both plastic enough to adapt to local needs and the constraints of the several parties employing them, yet robust enough to maintain a common identity across sites' (ibid.:393).

In a similar way to boundary objects, informed consent might be understood in terms of the tensions of plasticity and robustness when comparing dominant conceptual theories and their final implementation. In other words, when informed consent reaches the actual patient–researcher interaction, we its conceptual plasticity turns into an effective robustness. Here, informed consent seems to function almost automatically. The patients and the researcher implicitly know how they will deal with or how they will respond to it. Both share one specific interest—the patient's surgical intervention. However, their motives differ significantly: while patients undergo surgery *as a treatment*, the researcher depends on surgery *as a source of research material.* During informed consent both are able to realize their interests in the hospital setting. The researcher's strategy is to lead the informed consent conversation and to guide the patients' reading of the consent form. The patients react with the strategy of judging what they are told according to their attempts to contribute best to the preparations for surgery. The consent form serves all those purposes more as a standardized tool that enables interaction across divergent viewpoints than as a document that guarantees an 'objective' disclosure of information. In the end, consent is attained because the different worlds of the patient and the researcher have been successfully integrated while their individual priorities remain unchallenged.

17.6 **Implications**

As has been shown elsewhere (Felt et al. 2009), the concept of donors' individual autonomy does not adequately explain the decision practices during informed consent. Although a number of qualitative empirical studies have pointed out that informed consent procedures need to be understood as context-dependent interactions, most of the medical sector and its concomitant ethical review boards continue to frame informed consent as either safeguarding the autonomy of subjects or protecting researchers and research units from potential litigation. This view of consent also persists in the initiatives to build international and global networks of research biobanks in which donor consent is recommended, but is not obligatory, and priority is given to national legal regulations regarding consent requirements (OECD 2007; Zatloukal and Yuille 2007). Hence current practices of informed consent are reinforced rather than questioned, and the debates around consent are closed down rather than encouraged.

In-depth analysis of the concerns of the concerns of patients and researchers concerns during informed consent to research on leftover tissue leads to the conclusion that engaging in informed consent serves both parties in a number of ways. Both benefit from including consent in their respective activities. Thus, treating informed consent as a request intuitively bridges the different motivations of patients and researchers. Consequently, the donor's signature on the consent form takes on specific meanings that might not easily be interpreted as giving approval to the researcher's scientific project.

Therefore consent interactions need to be understood as institutional procedures that include a variety of purposes on each side—on the side of the patient-donor as much as on the side of the physician-researcher, to name just two. As any donation of bodily substances is likely to be involved with the health-care system and its professionals in some way, similar results have been reported even for cases where donation is not connected with any immediate health service, for example voluntary blood donation to a blood bank (Dalsgaard 2007). Even though blood donation involves insertion of a needle and the risk of fainting, donors and employees at the blood bank view it as exchanging an excess substance. This corresponds to how patient-donors of leftover tissue consider the removed tissue to be 'waste' or 'surplus'. As has been shown by Svendsen and Koch (2008), classifying human substances as 'spare' and simultaneously as legitimate sources for research involves considerable social negotiation processes among professional actors. Those negotiations neither stop at the bedside nor pause during informed consent. Approaching consent interactions as purposive institutional procedures might shed further light on how the actors involved come to attach meaning to the flow of human substances between the conflicting demands of research and care.

References

Berg, J.W., Appelbaum, P.S., Lidz, C.W., and Parker, L.S. (2001). *Informed consent: legal theory and clinical practice.* New York: Oxford University Press.

Bister, M.D. (2009). 'Jemand kommt zu Dir und sagt bitte.' Eine empirische Studie zur Gewebespende im Krankenhauskontext. (An empirical study of tissue donation in hospitals). *Österreichische Zeitschrift für Soziologie*, **2**, 72–8.

Black, J. (1998). Regulation as facilitation: negotiating the genetic revolution. *Modern Law Review*, **61**, 621–60.

Bourdieu, P. (2001). Soziales Sein, Zeit und Sinn des Daseins (Social being, time and meaning of existence). In P. Bourdieu (ed.), *Meditationen: Zur Kritik der scholastischen Vernunft*, pp. 265–315. Frankfurt am Main: Suhrkamp.

Busby, H. (2004). Blood donation for genetic research: what can we learn from donors' narratives? In R. Tutton and O. Corrigan (eds), *Genetic databases: socio-ethical issues in the collection and use of DNA*, pp. 39–56. London: Routledge.

Corrigan, O. (2003). Empty ethics: the problem with informed consent. *Sociology of Health & Illness*, **25**, 768–92.

Dalsgaard, S. (2007). 'I do it for the chocolate': an anthropological study of blood donation in Denmark. *Distinktion*, **14**, 101–17.

Dixon-Woods, M., Williams, S.J., Jackson, C.B., et al. (2006). Why do women consent to surgery, even when they do not want to? An interactionist and Bourdieusian analysis. *Social Science & Medicine*, **62**, 2742–53.

Dixon-Woods, M., Ashcroft, R.E., Jackson, C.J., et al. (2007). Beyond 'misunderstanding': Written information and decisions about taking part in a genetic epidemiology study. *Social Science & Medicine*, **65**, 2212–22.

Faden, R.R. and Beauchamp, T.L. (1986). *A history and theory of informed consent.* New York: Oxford University Press.

Felt, U., Bister, M.D., Strassnig, M., and Wagner, U. (2009). Refusing the information paradigm: informed consent, medical research, and patient participation. *Health*, **13**, 87–106.

Gimlin, D. (2007). Accounting for cosmetic surgery in the USA and Great Britain: a cross-cultural analysis of women's narratives. *Body & Society*, **13**, 41–60.

Hoeyer, K. (2008). The ethics of research biobanking: a critical review of the literature. *Biotechnology and Genetic Engineering Reviews*, **25**, 429–52.

Hoeyer, K. (2009). Informed consent: the making of a ubiquitous rule in medical practice. *Organization*, **16**, 267–88.

Hoeyer, K. and Tutton, R. (2005). 'Ethics was here': studying the language-games of ethics in the case of UK Biobank. *Critical Public Health*, **15**, 385–97.

Hoeyer, K. and Lynöe, N. (2006). Motivating donors to genetic research? Anthropological reasons to rethink the role of informed consent. *Medicine, Health Care and Philosophy*, **9**, 13–23.

Kopetzki, C. (2004). Die Verwendung menschlicher Körpersubstanzen zu Forschungszwecken (The use of human body substances for research purposes). In C. Grafl and U. Medigovic (eds), *Festschrift für Manfred Burgstaller zum 65. Geburtstag*, pp. 601–17. Wien: Neuer Wissenschaftlicher Verlag.

Mayring, P. (1997). *Qualitative Inhaltsanalyse. Grundlagen und Techniken* (Qualitative content analysis. Principles and techniques). Weinheim: Deutscher Studien Verlag.

OECD (2007). *OECD best practice guidelines for biological resource centres.* Available at: http://www.oecd.org/dataoecd/7/13/38777417.pdf (accessed 9 June 2008).

Star, S.L. and Griesemer, J.R. (1989). Institutional ecology, 'translations' and boundary objects: amateurs and professionals in Berkeley's Museum of Vertebrate Zoology, 1907–39. *Social Studies of Science*, **19**, 387–420.

Strauss, A. and Corbin, J. (1990). *Basics of qualitative research: grounded theory procedures and techniques.* London: Sage.

Svendsen, M.N. and Koch, L. (2008). Unpacking the 'spare embryo': facilitating stem cell research in a moral landscape. *Social Studies of Science*, **38**, 93–110.

Yuille, M., van Ommen, G.-J., Bréchot, C., et al. (2007). Biobanking for Europe. *Briefings in Bioinformatics*, **9**, 14–24.

Zatloukal, K. and Yuille, M. (2007). *Information on the proposal for European Research Infrastructure: European Bio-Banking and Biomolecular Resources.* Available at: http://www.biobanks.eu/index.html (accessed 9 June 2008).

Epilogue

The chapters in this book demonstrate very clearly the decisive question regarding the regulation of tissue research and biobanking at the current time. How is it possible to bring together fundamental aspects of ethics (human dignity, respect, trust in science and society) and law (property in body material, commodification and commercialization of the human body) with pragmatic regulatory aspects to enable the effective planning and performance of research projects?

This book shows the wide range of activities which were, and still are, permitted in European cultures and the traditions of scientific research with tissues and human body parts of living and deceased individuals. The interpretation and analysis of terms like 'human dignity' and 'respect for persons' also shows a new currency for older considerations. For example, the Kantian definition of not treating a person as a mere means gains new weight in the field of human tissue research, where body parts *are* usually used as a mere means for research and scientific progress. As we have seen in this volume, this is also true for the principle of non-commercialization. Although it is a cornerstone of the perception of the human body, this does not prevent the economization of human tissues and cells during their many applications in research. Therefore such recent definitions and considerations are clearly helpful for up-to-date reflections on changes concerning our traditional concepts and dealings with the human body in the context of scientific research.

The connection between solidarity as a normative principle, on the one hand, and health care as a particularly sensitive area of human endeavour requiring ethical and legal regulation, on the other hand, is demonstrated in the area of research. Research with human tissue, especially epidemiological research, shows the neeed to analyse our previous understanding of solidarity in health care and to try to develop new ways of integrating the concept of solidarity into medical research.

The examples of national regulations and jurisdictions provided in this book demonstrate that in many European countries it is seen as essential to develop detailed regulations to govern new approaches to research. The examples also show that nowadays an interesting and often well-functioning cooperation between the broader public, professional bodies and institutions, national ethics commissions, parliaments, and governments exists at the national level. How this work on the national level could be extended to the supranational level of the European Union and the Council of Europe remains a fascinating question for the future. However, a closer examination of the national frameworks for tissue and biobank research reveals differing levels of regulation. For example, at present there is no uniform requirement or standard for the national organization of an independent oversight for population biobanks, although this was recommended in Article 19 of the Recommendation document by the Council of Europe's Committee of Ministers.

An important point in this context of more uniform regulation is the question of the relationship between new and existing tissue collections. Should (molecular, genetic) research be undertaken with such collections, and under what circumstances? Human tissue research cannot be understood without looking at neighbouring fields such as organ donation or tissue engineering.

Keep developments in both therapy and research in mind, to avoid diverging standards, will be an important task for the future.

The ethical and legal regulation of medical research as a whole is predicated on the interests and rights of the individual patient who must be protected against the demands of research in particular and undue societal obligations in general. This perspective results, inter alia, essentially from the history of the abuse of patients and disadvantaged groups for the aims and goals of research and scientific progress. This abuse is best illustrated in the field of tissue research by the UK organ retention scandals, which showed fundamental differences in the perception of the appropriate handling of organs and tissue between researchers, physicians, patients, patients' relatives, and the broader public. However, recent developments in tissue research including, for example, large-scale epidemiological biobanks show that the individual perspective has to be balanced or complemented by considerations concerning the overall public good (an aspect which can be found in most European constitutions). The persuasiveness of this argument also depends on some empirical and strategic considerations, i.e. whether it is plausible and desirable that the aims of this branch of research branch can be achieved. From our point of view, the task of balancing individual interests against societal interests or the public good also explains reflections on whether the regulation of tissue research should be relaxed to create a practicable regulatory framework for such epidemiological endeavours. A good example of such efforts is the regulation of informed consent. In the context of biobank research, there are a number of considerations regarding the necessity of informed consent as well as the extent of the information that should be given to the individual participants (the discussion of broad or narrow consent). It is not surprising that those countries with a positive tradition of epidemiological and population research, such as the UK and the Scandinavian countries, find it easier to address the regulatory demands of tissue research and biobanking. The future will show whether such a positive perception of epidemiological research will also have an influence in other European countries.

A very interesting and unusual feature of tissue research and biobanking is that this kind of research normally requires the cooperation of researchers from different countries, research cultures, and disciplines who are subject to similar, but often diverging, national regulations. Many of these countries do not have a single comprehensive regulation or item of legislation concerning research with human subjects in general or tissue research in particular. Often, the relevant regulations and legislations are fragmented in drug laws, guidelines for researchers and physicians, data protection laws, guidelines for the security and safety of therapy and research, and even funeral laws, which have to be interpreted and underpinned by jurisprudence and reference to national constitutions. Interestingly, the development of medical ethics with documents such as the Declaration of Helsinki, and also with supranational initiatives and documents such as the European Convention on Human Rights and Biomedicine (Oviedo Convention), increasingly serve as a common framework and a point of reference for national ethical and legal developments.

Therefore the question arises as to whether the European Union could or should become active and establish common regulation in areas such as health, science, and research, which are thus largely seen as falling into the regulatory competencies of individual member states or even regions. However, it would be fatal if such a process were to be perceived as an enforcement of ethical and legal regulations against the interests of individual member states. The most promising alternative to such a centralized top-down approach seems to be the well-structured ethical, legal, and scientific discourse pursuant to a bottom-up-approach. Such a discourse could identify the most important similarities and differences in the various countries as an aid to obtaining a common political solution. An appropriate method for such a systematic undertaking for the

27 member states of the European Union is currently available but should to be radically improved in the future.

There are excellent arguments for a joint regulatory approach in the European Union—the advantages to research and cooperative projects in the sciences, and also the homogenous protection of the interests of the patients and citizens of Europe. The price which may have to be paid is of the abandonment or restriction of national or local traditions in research, ethics, and law.

Index